Front

Co-Chaos Patterns

The Universal Fractal

Katya Walter, Ph.D.

-⋄⋄⋄-

**The author thanks the
Institute for Neuroscience and Consciousness Studies
of Austin, Texas
for its helpful support and encouragement**

-⋄⋄⋄-

THIS BOOK IS
VOLUME 2, FOURTH EDITION
IN THE
TOUCHING GOD'S TOE SERIES

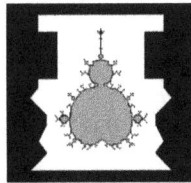

TO VISIT THE DOUBLE BUBBLE UNIVERSE, GO TO......
https://www.katyawalter.com

TO VISIT KATYA WALTER'S YOUTUBE CHANNEL, GO TO...

KatyaWalterYouTube
Kairos Center Publications
Box 142086
Austin, Texas 78714
kairospublications@gmail.com

Co-Chaos Patterns: The Universal Fractal by Katya Walter, Ph.D.
Volume 2 in the *Touching God's TOE series*, 4th edition
Copyright © 2004 by Kairos Center; 4th edition by Kairos Center
Editor: Jennie Rosenblum   Art by Adele Aldridge, Adrian Frye, & Katya Walter
                                  - or from Wiki Commons or Creative Commons

Paperback 4th edition published 2019   ISBN 978-1-884178-51-1
Electronic 4th edition published 2019   ISBN 978-1-884178-49-8

Library of Congress Cataloging-in-Publication Data
Walter, Katya - *Co-Chaos Patterns: The Universal Fractal*
Includes table of contents, appendix, bibliography, & illustrations
  1. Physics—gravity, cosmology, strings, spacetime, dimensions, fractal topology
  2. Religion—Touching God's TOE, spirit & science, religions, divine love
  3. Gravity—gravitation, unification, forces, emergent properties
  4. Chaos Theory—Lorenz attractor, fractals, chaos patterning, complexity
  5. Philosophy—Plato, Taoism, Chinese thought
  6. Mathematics—nonlinear, analinear, fractals, analog & linear number
  7. Mysticism—mystic love, remote viewing, dreams, I Ching, synchronicity systems
  8. Title: *Co-Chaos Patterns: The Universal Fractal*

-·:·-

Katya Walter's Books
Chaosforschung (in German) - *1992 - Diederichs Verlag*
Dream Mail: Secret Letters for your Soul - 1995 - Kairos Center
Tao of Chaos: Merging East and West - 1994 - Kairos Center - This original book was split and augmented to become Volumes 2 and 3 of the *Touching God's TOE series*, first published in 2004, and updated in a 4th edition, as shown below.

Touching God's TOE series, 4th Edition

Vol. 1: Double Bubble Universe: The Paradigm	*2018*
Vol. 2: Co-Chaos Patterns: The Universal Fractal	*2019*
Vol. 3: Tao of Life: The Fractal Gift	*2019*
Vol. 4: The Universe Is Alive and Well: The Organism	*2020*
Vol. 5: Master Code Tree: The Expansions	*2022*
Vol. 6: Stone Soup Universe: The Hologram	*2023*
Vol. 7: The Particle Ark (projected)	
Vol. 8: Quantum Organics (projected)	

Table of Contents

Initial German Review

The German publication of an initial single volume sparked this series. It was published in German as **Chaosforschung** *by Diederichs Verlag. Claus Claussen wrote this review that appeared in the magazine* **Neues Denken und Handeln** *in November 1992. That original book was later split and amplified into Volumes 2 and 3 of this series, so at the front of both volumes appears this translated review. [Permission was given to adapt the next-to-last paragraph slightly to fit the larger scope of the whole series.]*

"*Universal Life Pattern* could be a subtitle for this lofty theme that will pique your interest in the Orient. It might also be called *Breaking a Universal Code*, because it opens the door on a fascinating view of life. Number, more exactly, archetypal number, is the key to this research on chaos theory, Chinese philosophy, and DNA.

"Katya Walter, prominent philosopher from Texas, a Ph.D. who also has studied at the Jung Institute in Zurich and taught for a year at Jinan University of Canton, goes to the source of life's dynamic pattern in her book. She describes how the DNA spiral of our linear-minded Western science relates to the analog-style thinking of the old I Ching. She shows that the genetic code and I Ching function through the same chaos patterns, and that the physical system of DNA can be translated mathematically into the psychic system of the I Ching.

"Other scientists, and especially Martin Schönberger (1973) in his book *Verborgener Schlüssel zum Leben—Weltformel I Ging im genetischen Code,* have earlier pointed out an astonishing correspondence between the genetic code and the I Ching.

Walter makes reference to this work, but adds a new analog perspective, even enlightenment beyond Schönberger's book, going deeper and wider. Very concretely and beyond speculation, she lays bare a decodable correlation between amino acids and hexagrams. She shows that biochemical laws and old wisdom are connected through this mathematical pattern. It garbs old Eastern truth in new Western clothing. This chaos supersystem is provable with new terminology and computer graphics.

"Threading through the awesome labyrinth of this stunning theme, your guide Katya Walter continually startles you back into clarity with her personal engagement in the search for truth. She gives sidelong glances into her dreams, talks of her experiences and frustrations, and even jokes along the path. At such times her tone, normally scientific and yet crisp with a refreshing simplicity, takes on a more poetic lilt.

The author takes an informative stroll through the chaos garden as she explores its profound central theme, approaching it from three distinct vistas: I Ching, chaos theory, and genetic code. This sight-seeing tour is designed to render each path fascinating yet familiar. Otherwise the waves of scientific proof could become too big.

"Above all, this carefully crafted work is a treasure trove chock full of jewels. Finally, there is a special paradoxical treasure at the bottom of the chest: without ever leaving the groundwork of science, it moves beyond logic into universal values."

Claus Claussen

> To see a World in a Grain of Sand
> And a Heaven in a Wild Flower
> Hold Infinity in the palm of your hand
> And Eternity in an hour….
>
> *William Blake*

Introduction–What Is this Book?

From the Author: Astonishing parallels exist between the genetic code and ancient China's I Ching math figures. Viewed together, both systems hint at a master code that generated the universe itself. All three systems, taken together, offer us a decodable Rosetta Stone whose shared, easy math can decipher the underlying master code that generates space, time, matter, and energy.

For a specific range of numbers, this math is nonlinear in a sense so special that I call it *analinear*. To a math whiz, its explanation may even be tedious. But I write for the ordinary person who wants to understand how the code works.

The I Ching was said to "show the way of the Tao." Breakthroughs in modern science are now slowly rediscovering this fundamental code, but doing it piecemeal, in various labs and diverse theories, and in mathematical study itself. In this volume, we examine the I Ching's yin-yang structure grown on a dp-tree, and we find that it is both fractal and analinear. We detect many physical clues, philosophical parallels, and odd synchronicities that offer stepping stones to understanding the universe itself.

In 1985, I began writing an initial book, *Chaosforschung (Tao of Chaos)*, that morphed into this series. Overwhelmed by the data, I went to sleep one night asking for a dream that would show me how to lay it out appropriately.

I had a dream…

I'm in a broad, beautiful land with many trees. It is night. In the center of the forest, I look up at a huge old tree, dark against the starry sky in its detail of branch and twig. There's room enough for all of us here in this big, intricately textured garden, I realize. But some want to cut down the trees and level it out so that huge throngs of people can gather to gaze up at the sun's glare.

I watch dark twigs fingering the remote stars.

A voice speaks: "Don't turn this into a Copernicus Garden."

The next morning, I woke up and sleepily wrote down the dream, pausing to ponder that name I'd heard: *Copernicus Garden*. Hmm. It set me to thinking about how Copernicus put the bright light of logic at the center of our science, mostly by proposing that the sun is at the center of our solar system. That feat by Copernicus led us to appreciate the scientific method.

However, the dark garden in my dream offered a starry night sky showing not just one sun, but many suns twinkling through the branches of a huge old tree. That big tree recalled to me the Tree of Life, which surely holds in its branches many living beings in many solar systems of many stars.

Due to that dream, I decided not to lay out this series of books in a purely scientific format that might minimize the wondrous old tree of life at its center.

FOR AN OVERVIEW OF THE WHOLE SERIES, LOOK FOR THE SERIES SUMMARY AT THE BACK.

I won't forget the many stars in its branches that await us beyond our sun.

This book is a dialogue between you and me, a conversation as we stroll in a moonlit, fractal garden where twigs finger the stars. Here creation nests in scaling patterns that have a self-similar yet never quite repeating beauty. Here we can contemplate webby connections of thought that merge into patterned insight.

No matter how huge, this garden stays life-sized because we have a place in it, you and I. There is no need to cut down the layering branches, no need to cart off the trees and let the bright Sol of left-brain logic, our sole star of daytime dazzle, become so glaring that it blinds us to the right-brain's holistic appreciation of myriad other stars glowing in the constellations beyond.

In this fractal garden, there is room enough for bright dreams and dark doubts, for quick discoveries and slow evolution. There is room to grow. Thus we needn't be perfect, or failing that, be cast out. Perfection is end-stopped, but this fractal garden allows change. This garden is not perfect, but it is the perfect place to learn, grow, and evolve. Here we can walk with divine nature that is visible everywhere, yet finally unknowable. Its majesty stretches beyond our human ken into darkness, yet it willingly shares as much as we can bear to see.

From the Editor: This book is Volume 2 in the dazzling *Touching God's TOE* series, 4th edition. This series shows how the I Ching and DNA are two known variants of a hidden master code that generated the universe. Examining parallels in all three codes offers us a Rosetta Stone to decipher the master code.

Called in Germany a "philosopher queen of the global village," Katya Walter, Ph.D., in this volume sets out the fractal foundation for the master code. She shows how it began not only as polarized binary units (0, 1), but also as polarized analogs bifurcating from the zero of nothing (0 = -1 and +1). She explores how co-chaos patterns in polarized pulsing generated our universe. Along with scientific concepts, she also examines philosophical parallels and recounts events inspiring the series.

This series began as one volume, *Chaosforschung*, published in German in 1992, and in English as *Tao of Chaos* in 1994. It was later split and amplified into Volumes 2 and 3 of the series. Its odd-numbered chapters (1, 3, 5...) are more science-based. Even-numbered chapters (2, 4, 6...) are more philosophical and personal.

This book has 19 chapters and a *Series Summary,* comprising 105 sections. It has a *Bibliography* and *Reviews.* It has 70 listed images, graphics, and charts. The ebook version also has an interactive table of contents and 87 e-links that act as informative footnotes. Its text is completely searchable and receives electronic updates. It is also hand-edited to hold color graphics that allow greater distinctions in images and charts. Consider getting both the print and ebook versions for a greater range of information and versatility.

Each book in this series has its own symbol. The symbol for Volume 2 is the fractal hyphen (-) from *Fractal Font* by Tibor Lantos. It is much enlarged here.

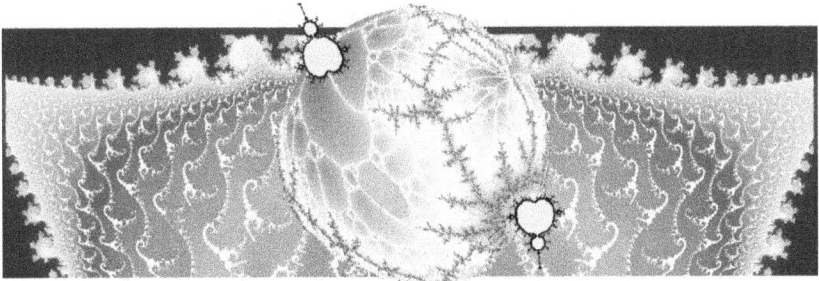

Chapter 1: Analinear Co-chaos & 20 Big Questions

1. Let's go exploring

On March 4, 1985, I had a great dream that showed me the layout of our Double Bubble universe and how the four primals of space, time, matter, and energy developed their structure. It was magnificent and almost unbearably beautiful, but it was also quite incomprehensible to me, logically speaking.

I did not have enough physics or math background to understand the layout, structure, and dynamics that I both *saw/was* in some mirror-reversed fashion during the dream. I just experienced it wordlessly, like a babe lost in the wild woods. I gazed around in wonder and bewilderment, shocked and puzzled at the universe's simple brilliance and strange inevitability.

Even more baffling in the dream, it was *God* who showed me that majesty in universal nature! Me, an agnostic! Why would such a dream come to me? At that time, I was teaching in the English department at the University of Texas at Austin. But when summer came and my husband retired, I wanted to go study at the Carl Jung Institute in Zurich. Why? I needed to understand the bewildering, magnificent dream and why it had come to me.

Fortunately, John happily agreed to go to Switzerland with me, so I quit my job, and for the next 5 years, I studied at the Jung Institute and in the library of Zurich's Eidgenossische Technische Hochschule, trying to understand what I might, could, must do about that astounding dream. It seemed like both a gift and a responsibility to tend, for why else would such an amazingly complete origin of universal structure be presented to scientifically-naive me?

That's how this quest began. I soon realized that I'm not alone in my naiveté regarding some key concepts about the universe. In his Introduction for a 2007 *Science* magazine article, "Quarks and the Cosmos," cosmologist Michael Turner said of the universe, "Although we know much about the universe, we understand far less." Over the next 35 years, I gradually came up with 20 questions and their answers for what I'd witnessed in that great dream.

Now allow yourself to explore the workings of the remarkable Double Bubble universe with the simple curiosity of an inquiring child.

2. Two ways to measure time

We embark on this description with an example from everyday life. It's a description of two ways that we measure time.

Analog Face   Digital Face

Two views of time

Consider these two watches. They both show the same time: 9:30. The watch on the left has a round, analog face. It shows time as a cycling, continuous rhythm. The watch on the right has a square face. It shows time as discrete, unitized jumps of successive digits.

The round watch face reveals that timekeeping has related cycles. As a child, you learned how the *second* hand circles the dial in 60 seconds to mark a minute. The *minute* hand circles the dial in 60 minutes to mark an hour. But the *hour* hand has to circle the dial twice to mark all 24 hours of one day and night.

Watching both hands go around on the moon-faced dial gives a child an overall sense of the seconds, minutes, hours, and days in cycling relationship. They all make a circling motion, and each measurement relates to the others.

On the digital watch face, however, you just notice a series of numbers flicking by. They leap on and off the face, chopping time into chunks without telling you their proportion to anything. And although the watch is called *digital*, it doesn't use the base-10 system that came from counting on our ten fingers. We don't expect the numbers on that square face to jump from 9:59 to 9:60, 9:61, and so on, rising higher and higher to reach 10:00.

Yet in a purist sense, shouldn't a *digital* watch employ the same digital counting sequence used by the metric system, most money, the Dewey decimal system, and so on? A truly digital watch would count off 100 seconds in a minute, 100 minutes in an hour, 100 hours in a day. Logically speaking, that is.

Face it, digital time is deceptive. Glancing at a digital watch won't show you the overall proportions of our cycling 60-second, 60-minute, 24-hour day and night. Yet proportional relationships in numbers really can matter.

That's why most airplanes today no longer use a "modernized" digital readout for the fuel supply. Manufacturers went back to installing the older, analog face. A digital readout for the gas tank only conveys raw numbers, not the actual proportions involved. If you've got 40 gallons of fuel left, is that tank almost empty? Or almost full? An analog dial will immediately tell you the vital difference, but a digital readout does not.

3. Analogies open up analog thinking

There are limits to logical, linear thinking. Sometimes something can make good sense even when it does not make *logical* sense. This happens constantly in ordinary life. Jokes, for instance, are not logical, yet they make a weird kind of sense. A joke usually goes through several cycles of escalating intensity until an abrupt twist suddenly pulls you up short…and laughing.

You can easily discern a funny joke from an unfunny joke—in other words, you make a quick judgment call on its *quality*. And this happens not only with jokes. With friends, enemies, or loved ones, you may experience an intuitive insight that's beyond ordinary logic. Sometimes events will contain a gut-level patterning and coherence that you cannot explain with cause-and-effect logic. But maybe you can explain it with an analogy.

Analogies are approximations. They hide in remarks like "What goes around comes around." "It takes two to tango." "That'll cost an arm and a leg!" "I heard it on the grapevine." "You're flogging a dead horse." "He threw his hat in the ring." "Know when to hold 'em, and know when to fold 'em." All of these idioms suggest implicit relationships of one kind or another that vibrate with associative meanings. You can easily understand such talk, even if you cannot grasp it exactly by the rigid handle of linear logic.

Analogies are often simple. They are intuitive. They make comparisons. They use parallel relationships to spotlight a similarity or a difference—often evident, sometimes quixotic. An analogy brings up all sorts of hinted connotations and resonant implications. It rings in associations that can imbue an event with energetic force that is outside straight-line logic. Echoing vibes shimmy and shiver in an analogy, impalpable yet potent. They open the door to an ongoing process rather than closing it down in a definite, summary *yes* or *no*.

The process that's so evident in analogies can even simplify math. Philip Davis and Reuben Hersh say in *The Mathematical Experience,* "Analog mathematizing is sometimes easy, can be accomplished rapidly, and may make use of none, or very few, of the abstract symbol structures of 'school' mathematics. Results may be expressed not in words but in 'understanding,' 'intuition,' or 'feeling.'"

Polish mathematician Stefan Banach also touted the utility of analogies in math: "A mathematician is a person who can find analogies between theorems. A better mathematician is one who can see analogies between proofs. And the best mathematician can notice analogies between theories."

4. The pair of pairs turn into algebra!

An analogy that is spoken in a sentence can hold hidden math. For instance, take the evening my friend Ted said: **"Marilyn was to the 1950s what Elvis was to the 1960s."** To Euclid or Euler, any meaning or math would not be obvious. Marilyn who? What went on where? And what's an elvis?

"How do you mean?" I asked.

Ted made swift leaps through pop culture to illustrate his point. Here's the short version: Marilyn Monroe was a female movie star of the 1950s. Elvis Presley was a male rock star of the 1960s. Both sprang from poverty. Both lived glamorously. Both rose to pop Olympus fame. Both were voraciously vulnerable sex symbols. Both died tragically in dramatic circumstances. Both became cult figures. If you know their history and mystique, you catch the drift of Ted's analogy right away. It's all hidden in there.

Notice, this analogy is not just a comparison between two things, but among four things…a pair of pairs. Ted compared two different stars in two different decades, and he implied parallels within those conditions. He described a relationship between a pair of pairs, but it's only approximate.

We can put Ted's free-wheeling analogy into a mathematical form. Why bother? Because it is key to understanding the I Ching's math shorthand!

(*a*) *Marilyn* was to (*b*) the *1950s about like* (*c*) *Elvis* was to (*d*) the *1960s*

$$a \text{ is to } b \approx c \text{ is to } d$$

Ted's remark is suddenly algebra, where ≈ means *about like* or *approximately* or *not exactly equal.* Approximation turns straightforward = into wavy ≈ !

The circles below also contain a hidden example of a relational pair of pairs. In Euclid's plane geometry, you can make circles of any size, anywhere, as often as you want. But any circle will always have the same ratio of circumference to diameter. Cultures worldwide have honored this ratio. Math symbolizes it as π (called *pi*), the sixteenth letter of the Greek alphabet. The π symbol describes the only way to fit a circle together, no matter how many soap bubbles you blow. And the fractal Mandelbrot set cannot iterate without using π.

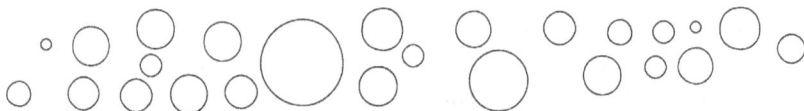

Circles

A hidden pair of pairs exist inside the meaning of π. This can be expressed by numbers, so let's do it. A circle's circumference is to its diameter *about like* 22 is to 7. Thus the value of π is *approximately* $^{22}/_7$. In this way, π sets up a mathematical analogy using a relational pair of pairs. So...

    1. (*a*) **circumference** is to (*b*) **diameter** *about like* (*c*) **22** is to (*d*) **7**

    2. *a* is to *b* ≈ *c* is to *d*

    3. *a* : *b* ≈ *c* : d

    4. *a*/*b* ≈ c/d

Above are four different ways to describe the relationships of π's hidden pair of pairs. In each version, all four parts relate to each other...approximately.

The ratio of $^{22}/_7$ is only an approximation of π, so here is where the analog process develops an elusive magic in its relational heart. To get a bit closer to an exact numerical ratio of π, it is more nearly about like $^{355}/_{113}$. But π can never be set into a truly exact ratio. In other words, π cannot be expressed precisely by any ratio or numbers. Since it is not truly ratio-able, π is an *irrational* number.

And if we do try to force π's inexact ratio of approximately $^{22}/_7$ into exact units of linear number, woe be unto us! I can say succinctly, correctly, that the ratio of π is approximately $^{22}/_7$, but if I then divide 22 by 7 to get a more exact number, my answer will start with "π is 3.14159265358979323846..." and go on and on. Those numbers just keep extending. Forever. They continue to literal infinity. Talk about *pi* in the sky!

By now, π has been calculated to trillions of decimal places, but as physicist Paul Davies said, "...decimal places of π form a completely erratic sequence." And unless you memorize a lot of that sequence, you cannot even predict which number is coming up next in the infinite string that starts with 3.14....

We cannot say that π is just random numbers, though, since its digits are fixed. For instance, the second decimal place in π is always 4. And yes, there is an algorithm, the BBP formula, that can compute a single specific digit without calculating preceding digits, but even that formula cannot find a final digit.

Thus, from π's relational proportions drawn in a simple circle, we have somehow stumbled into an endless sequence of ever-extending numbers. This happens whenever we try to turn the fraction of *just about* $^{22}/_7$ into an exact, final answer. Its succession of ever-extending numbers will keep on trying fruitlessly to express π's proportions in precise, unitized digits that plod toward a summary goal of *The End*. But there is no end.

5. Goal-driven vs. process-oriented

When adding things up, numbers like to handle standard units of... whatever...driving to a goal. An instance is "2 sheep + 2 sheep = 4 sheep."

We're counting sheep here while going for the goal, the solution, the concrete answer to "How many sheep?" The numbers merely stand in for space-filling units of some sort—and here it happens to be sheep, the contents being carried in their number containers of $2 + 2 = 4$. We can also assume the answer will not somehow veer away from its additive goal. The answer won't be sheep^2.

However, some mathematicians are now beginning to avoid the = sign, instead favoring a looser relationship of "equivalence" promoted, for instance, by mathematician Jacob Lurie. His work seems to look beyond *quantity* in units to the *qualities* involved in algebraic topology relationships.

Numbers used in an analog way do not emphasize the *quantity* of units. Instead, they emphasize the *quality* of relationships. They compare ratios and proportions, where the leeway in *pi's* ratio of *just about* 22/7 can even resonate in modulations of partly. But if you try to force 22/7 to reach digital exactitude, a final answer, or even a predictable succession, its ongoing flow of numbers jeers, "I'm an unpredictable process without end!"

Yes, sometimes a flat, definite answer of *yes* or *no, this* or *that, 0* or *1* is useful and needed. But at other times, you may prefer to say *maybe, kind of, it all depends*. And if you teasingly say "Nooooo" to a friend, your sliding tone of voice makes all the difference in what you mean. You've just put a tailspin on *no* by giving it a double or triple message.

Such messaging puts the twist into analogies, either in words or in numbers. They are more connotative than denotative. They emphasize correlation, not cause and effect; quality, not quantity; process, not product. They engender stray resonances that linear logic does not want to encourage or deal with.

Traditional mathematical exactitude prefers to stick with uniform chunks in a linear sequence reaching a definite answer: if *this* and *this*, then *this*…so that adding up "2 sheep + 2 sheep does not result in the tailspin answer of sheep^2. Linear logic does not like to trigger a whole cascade of cycling resonances with unknown impacts and unexpected outcomes.

Yet analog processes often do just that—they trigger vibratory tailspin, whether you want it or not. It happens in an old movie when the soprano sings and a glass shatters on the table. Or when a mike's feedback starts squealing from loudspeakers. Or when you smile and your lover melts. Or when a football game's enthusiasm drives the crowd to leap up as a single, roaring beast, all vibrating together with the excitement of a touchdown. Or when the analog connectivity of mob psychology turns a throng into one huge, dangerous organism rioting outside the usual norms.

Analogs do not much care about the summary quantity of results. They're too busy exploring the quality of the trip along the way, making comparisons that energize the process, not striving for a goal so much as exploring the engrossing relationships that meanwhile keep unfolding...so that finally you may never even reach some goal over *there*. For analogs, *there* becomes irrelevant. *There* is no goal at all. Instead, you continue being *right here* in the process of traveling because, to quote a master of analogic, Gertrude Stein, "There isn't any *there* there."

Analogs turn us all into relatives, giving our human ties their amazing power to pull us along by the nose and wallet and heartstrings for years, for a lifetime. They send us hither and yon across the face of the earth. No, it is not logical that we fall in love or into despair or begin a quest for the Grail, but when it happens, an inexplicable power grips us, and our lives alter.

Research shows that cause-and-effect logic comes from the brain's left hemisphere, while analog holism is right-brained. (Okay, in right-handers... in lefties, it's mirror-reversed.) However, both sides of the brain work together more complexly than a mere left/right hemisphere division can insist on.

Nevertheless, the two sides of your brain in action serves as a useful metaphor for nonlinear processing. It is a microcosmic rendition of the larger mathematical principle that the universe itself is nonlinear, using linear, sequential units to go for the goal, as meanwhile, the analog, exponential variables keep searching out relationships in the shifting fit of things.

Fortunately, the amazing parallel processor that you carry around in your skull helps you fathom life's nonlinear events in a way that goes beyond simple logic. The two hemispheres of your brain provide an intimate proof that its two different approaches can inherently cooperate to make sense of reality.

Your intuition often spots a pattern that is submerged in apparent chaos. Small patterns nestle into bigger designs that are not logically evident, yet somehow they rise to your attention. This is the skill of pattern recognition... what you did as a child when you found six cats hidden in the drawing of a tree, although logic chides that six cats rarely hide in a tree.

I expect, for instance, that you are getting an intuitive feel for how this book develops its themes. A sudden *Aha!* and the light bulb turns on: "I get the drift here!" So you need not grasp every detail to get the main point. We'll ask some big questions, some of them beyond the limits of linear logic.

Know that in this book, numbers are not just cold stuff stacked in the sums of statistics. They are warm. They weave your flesh, your thoughts, the texture of your days. Here, you will find that numbers, normally considered coolly logical, leap to the intuitive heart of things beyond linear logic. Here, they dive into deep, qualitative values that our techno-culture neglects today, focused so long and hard as it has been on taking the discrete measure of things rather than noticing their qualitative relationships, intent on keeping an observer's cool, skeptical distance rather than acknowledging the intimate I-thou bond.

6. The 20 big questions

Current physics admits that it needs to find a Theory of Everything, a big TOE, to answer many puzzling questions that remain in our understanding of the universe. Volume 1, *Double Bubble Universe,* introduced 20 unsolved questions and proposed some answers for the first 5 questions.

This is the only other volume in the series that lists all 20 questions *en toto.* I try to keep the terminology simpler than you'd find in a normal science book. I use ordinary vocabulary (mostly). Why? In the first place, I'm not a scientist. Second, I write for all of us gathered here to talk at the table of life.

By the way, if you've already read Volume 1, *Double Bubble Universe,* you saw these questions there, so you may want to jump ahead to Chapter 2.

Question 1: What is our universe? Is it the only one? We have no physical proof of another. Physics studies the physical, material, mechanical, and energetic features that it can recognize and measure. It says our universe holds all the suns, planets, galaxies…all mattergy in all spacetime.

But this TOE suggests our universe is something more. It is organic and alive, although that's hard for us to recognize since the universe is so much larger than anything we usually define as a living organism. In fact, this TOE says we can understand the micro and macro levels of physics better if we turn the phrase *quantum mechanics* into *quantum organics,* and if we realize that our universe holds two huge, symbiotic bubbles with mirror-reversed properties.

Question 2: What is the working shape of the universe? Physicists wonder if it is round, open, closed, or flat? Maybe bouncing? This TOE says our Double Bubble universe is a gravitationally-flat hologram, meaning it balances out gravitationally. Why? Because its two bubbles hold counterbalancing poles of gravitation. Current science knows only our familiar bubble, and only what is visible to us up above the quantum scale. Our known bubble holds 3D space, an arrow of time, and a single gravitational pole.

But this TOE says that far below the *quantum* scale, where matter and energy emerge, is the ultra-tiny *mobic* scale, where space and time emerge. There a thin membrane of mobic pores (*mactors*) interfaces with a conjoined bubble, a hidden mirror-twin holding 3D time, an arrow of space, and the "lost" pole of gravity.

Light holography is already familiar to physics. It starts at the quantum scale, where matter and energy emerge. In light holography, a beam of coherent light can split and project two beams of photons that meet again in a 3D image.

But the universal hologram originates at the ultra-tiny mobic scale, where space and time emerge. That's where dimensional coherence splits, develops, and projects 3D space and 3D time, plus the potential for matter and energy. So photons are just one component in the Double Bubble hologram generated by splitting dimensionality at the mobic scale into two poles, space and time.

Question 3: How did the universe begin? Did it start from nothing? How do you get something from nothing? Why would it even bother to start? This series offers a possible solution to such questions. A simple and even rather amusing solution. Something even a 4-year-old can grasp.

Question 4: How deep in nature does fractal patterning go? Do fractals drive to the very root of things by generating the four primals that make our universal structure: space, time, matter, and energy? Or is fractal patterning just a frill on top, like a cherry on the ice cream sundae of life?

This series suggests that within nature's apparent randomness, fractal patterning hides at every scale of existence. It not only shapes our universe but even birthed it. This TOE says our universe began by developing a fractal master code that generated this holographic universe whose ongoing dynamic recalls a Lorenz attractor tracking events over two domains of reciprocal properties.

The master code eventually templated a minor variant: the genetic code that generated us smaller organisms in its upper-bubble gut. The genetic code that makes our bodies can help us understand the master code that made the universal body. Both codes use complementary chaos, or for short, *co-chaos*.

Question 5: Where did the original "lost" antimatter go? Physicists say when the universe began, it must have created about-equal amounts of matter and antimatter. Matter is still here. But the original antimatter went missing. Where did it go? Did it self-destruct? Get lost? Go hide?

This TOE suggests that much as Edgar Allen Poe's "Purloined Letter" was invisible to its oblivious searchers (a mantelpiece rack held the letter), so does the "lost" original antimatter hide below our conscious attention. It becomes evident, though, if we reframe our universal perspective enough to look below the mobic scale and see a conjoined, mirror-twin bubble. It has 3D time, an arrow of space, and the "lost" pole of gravity that attracted all original antimatter.

During creation, original antimatter was polarized to sort into the other bubble below the mobic scale. But that other bubble's skinny arrow of space crushed-converted its original antimatter into speedy tachyonic energy moving faster than the speed of light[2]. The super-swift energy constellated into the vast patterning of a huge universal mind spread throughout contiguous 3D time. It planned, birthed, curates, and cultivates us tiny organisms in its upper-bubble gut. We recognize its wisdom and call it Mother Nature. Occasionally we also tap more directly into that huge mind's data bank whenever we dream or go beyond the ego level in altered states that can tune into deeper realities.

-:⁝:-

The first 5 questions were examined in Volume 1, *Double Bubble Universe*. This book, *Co-Chaos Patterns*, will explore in more detail **Question 4: How deep in nature does fractal patterning go?**

You may ask, "Why spend so much time on just one question?" Doing so will help you comprehend in later books how the master code works to generate space and time, plus setting up the potential for matter and energy. We'll need a Rosetta stone to decipher it. First, we'll see how co-chaos works in the genetic code, then we'll show how I Ching math can shorthand its analinear dynamics. That gives us a Rosetta stone whose two known codes can help us unpack the master code.

We must establish the paradigm of co-chaos at the foundation of all three systems before we can grasp that our universe is not one bubble but two, and they sit inside of each other at opposite ends of the scaling spectrum. Such a paradoxical thought cannot be probed without employing the co-chaos paradigm.

-:⁝:-

Question 6: How did dimensions develop? Current physics does not know, but it recognizes that dimensions give us a way to perceive, describe, and use space and time. This TOE says that originating, polarized pulses of sheer *being* (vs. *nonbeing*) at the mobic scale generated space and time dimensionality.

Question 7: What is gravitation? Physicists think gravitation was the first force coming out of the cosmic egg. They say it is the only monopolar force. But some also wonder how gravitation can even work with just one pole?

Maybe something is off in the way that we conceive of gravitation? For instance, evidence shows us that galaxies spin too fast and their masses don't weigh enough to hang together gravitationally. Yet they *do* hang together. How? Physicists hypothesize an unknown *dark matter* (or perhaps even an invisible shadow universe) that provides the necessary hidden mass to hold all the galaxies together...but no one has yet found any dark matter.

Maybe a more fundamental factor is holding the galaxies together? Perhaps it is the universe's working shape? This TOE says our Double Bubble universe is a dynamical system that uses its two gravitational poles to hold itself together.

Question 8: Why does the universe seem to expand constantly and constantly ever faster? Physicists speculate that a strange, hypothetical *dark matter* holds the universe together. Dark matter would act the very opposite of an equally hypothetical *dark energy* that seems to be expanding space itself. However, no one has yet found any dark matter or dark energy.

Some even wonder if our notions of dark matter and dark energy may be due to a misunderstanding of the universal bookkeeping for gravitation as an emergent force. This TOE suggests that viewing the universe as a Double Bubble hologram made by separating gravitation's two poles can relieve scientists from fudging the books with mathematically devised hypotheses of dark matter and dark energy in what they assume to be a monopolar universe.

Question 9: Did the cosmic egg inflate in a hot Big Bang? Some physicists think a tiny originating singularity swiftly grew to quantum size. Then quicker than a wink, the super-hot dot instantly inflated to enormous size in a BIG BANG! Yet such a quick growth spurt also furrows their brows in puzzlement. They wonder how such bizarre, hot, instantaneous growth could happen? This Double Bubble TOE describes a cooler version of how such implausible growth might happen naturally, inevitably, organically.

Question 10: Why do so many equations have "nonsensical" reciprocal solutions that later turn out to reveal something important... perhaps a new particle or law? This TOE suggests there is no great mystery to the efficacy of reciprocal solutions if we recognize that they point to the Double Bubble universe's reciprocal symmetry. Suppose we posit that our known bubble has a hidden, reversing-mirror twin that is its complementary opposite. In that case, the symmetry of reciprocal equations becomes just a natural function of so much symmetry embedded at the deepest level of universal structure, and even in the master code that originated it.

Question 11: Why is physics plagued with "impossible" infinities in its measurements? General relativity indicates that matter and gravitation become infinite in the black hole at the center of our galaxy, and of each galaxy. Why? Moreover, at the other end of the physical scale, quantum mechanics says the tiny quantum level appears to hold an infinite amount of energy. Why?

At both extremes of the physical scale, scientists use calculations that provisionally rectify those "impossible answers" of infinity by just taking some zeros off the end. It is a sort of allowed cheating. But the Double Bubble TOE suggests an easier way to embrace those infinities at both ends of our bubble's sizing spectrum and reconcile them into a bigger, holographic picture.

Question 12: What is electromagnetism? What is polarity? We use electricity constantly in our technology, and even in our brains and nervous systems...but we may still misconstrue its essential character. This

TOE suggests that due to a simple switching error by Ben Franklin when he labeled electricity's + and − poles, science sometimes misconstrues the very nature of polarity. It has skewed our thinking on energy propagation and even on the basic polarity of electrons, photons, neutrinos, and especially of gravitation itself. For instance, why is electromagnetic force carried from one charged particle to another by a photon? Indeed, what is this weird little massless particle we call a photon? Electrically neutral and clone-like, a photon even acts as its own antiparticle. Why?

Question 13: Why does light in a vacuum move at an absolute constant speed? Why can't anything go faster? Does light's intrinsic gift reside in being a slow dobbin of constant virtue? Yes, says this TOE. The clone-like photon, both particle and antiparticle, acts as the benchmark in each bubble. It sets the means by which all matter and energy above and below the mobic scale get quantified, qualified, and brought into accord. We shall find that at creation, would-be antimatter tried to emerge inside the thin arrow of space in that other bubble as photons, but its potential got crush-converted into tachyonic energy that powered up a huge, unified mind in 3D time. People refer to this brilliant mind as Mother Nature, Akashic records, or Darwinian chance.

Question 14: Why won't Einstein's cosmological constant just go away? Einstein originally assumed this universe did not alter in size or destiny and was in fact eternal, so he put a stabilizing term into his theory of general relativity. He called his term a *cosmological constant*, and he used it to balance out gravitation and render it zero or gravitationally "flat" in an eternal universe.

Then Edwin Hubble overturned Einstein's idea of an eternal, static universe by showing that most galaxies appear to be red-shifting away from our own at ever-greater speeds. Embarrassed, Einstein called his own stabilizing term of a cosmological constant "my greatest blunder," and he recanted it.

But 45 years later, scientists began to wonder if some new version of Einstein's "blunder" might actually be correct? Could the once-derided cosmological constant again be re-employed to explain certain discrepancies in the standard model? This series says "Yes!" and it suggests a way—by putting a second, stabilizing bubble into the larger equation of the Double Bubble universe.

Question 15: Why can a particle-wave act like *either* a particle *or* a wave, depending on how it is studied in double-slit experiments? For a long time, scientists could not simultaneously and accurately measure both a particle's location and its wave speed. Moreover, the problem was not simply that particle-waves are too tiny to measure accurately. Instead, the accuracy of measuring a particle and its wave seemed to vary inversely.

So why does particle-wave identity seem to demand an *either-or* approach to measure it, not allowing a *both-and* embrace? After all, in the everyday world of Newtonian physics, it is easy for judges at the race track to track a horse's location and its speed at the same time. But for a particle-wave, it seemed impossible to measure both aspects simultaneously. Science needed a mind-shift.

Then, in 2015, a group at Ecole Polytechnique Federale de Lausanne managed to do it! This series describes how their approach made it possible to make the first-ever photograph of light as simultaneously both particle and wave.

Question 16: Can two particles communicate faster than the speed of light? Two distant but related or "entangled" particles can share information with each other instantly! Einstein died supposing that instant communication speed was impossible. But in 1964, John Stewart Bell theorized that two entangled particles can instantly influence each other without being in each other's locale. By now, Bell's theorem has been verified by many experiments. Particles really can "instant message" each other!

Wait! How can information go faster than the speed of light? Isn't it contrary to Einstein's theory of general relativity? Yes, such speed is theoretically impossible, but observably true, says current physics. No, it is naturally and inevitably true, says this TOE, and it explains how "instant messaging" occurs due to the tachyonic speed of lower-bubble data relayed to this upper bubble.

Question 17: What is the neutrino? Why is this hinky little particle so indecisive about its identity? Originating in our sun or other stars, the neutrino on its way to Earth oscillates through three different types of mix-and-match particle identity, each with its matter and antimatter versions. However, according to the standard model for particles, such oscillations are impossible!

Why? Because the rule says a particle can oscillate only if it has mass. You have to weigh to play. But the standard model allows no mass for a neutrino, although it acts like it has a tad. What trick is the neutrino playing on science?

Question 18: Is our universe designed to foster life? It is full of cosmological parameters that seem precisely supportive of life. Many scientists have noted this, but is it a mere coincidence somehow incredibly far from probability? Or do constraints in the universal design intentionally foster life within its cooperation of parameters? This Double Bubble TOE says our universe itself lives. It also fosters life within it, like us microbes exploring it with consciousness, and this series gives some instances and reasons why.

Question 19: How will our universe end? Or will it? Will expanding space eventually disperse all mattergy so much that it turns our universe into a cold, smooth, dark, entropic soup? Or will it crunch down into a hot dot again, reverting to what some scientists assume was its point of origin? Will it go out with a bang? Or a whimper? Will it reverse into a black hole? Or maybe undergo endless cycles of inflating boom and contracting bust?

None of the above says this Double Bubble TOE. Okay, then what? Our universe's future is implicit in its past, even in its origin. Upon its future death, due to our universe's inherent fractal makeup, it will likely iterate a few more variants as successive universes riffing upon its original co-chaos design.

Question 20: Is it possible to reconcile physics and philosophy in a TOE? This last question is so open-ended that I handle it differently from the others. In fact, **Questions 1** and **20** are bookend questions. I address them both throughout the series, often in tandem with other issues.

-:ξ:-

Did you ever wonder why reality doesn't crash like a glitched computer? This series of books says it's because our Double Bubble universe is a huge dynamical system of analinear co-chaos. It keeps this universe strong and enduring, safeguarded yet evolving, by means of a simply perfect algorithm.

The Double Bubble universe is, in fact, a living system, and it operates with so very many fractal failsafes that they manage to maintain the proportional fit and shimmy of our holographic reality in subtly flowing fashion.

Just as you do. Your own life is nonlinear, and it is nourished by the same dynamics. Fractal patterning shapes your body to keep its proportions in workable relationship. It makes your feet big enough to hold up your body, but not so big that the platform becomes too cumbersome to carry around easily.

Fractals shape a tree so that its branches do not uproot the trunk (usually) when its fluttering, green mass of leaves catch a big wind. In this material, graspable, measurable landscape of roads and buildings, bolts and levers, sea level and soap bubbles, the many chaos patterns operate in such a tenuous, shifting, intangible fashion that we experience life more than thinking it through. Meanwhile, the analinear supersystem even manages to protect our little ecosystem from overloading on too much human input...so far.

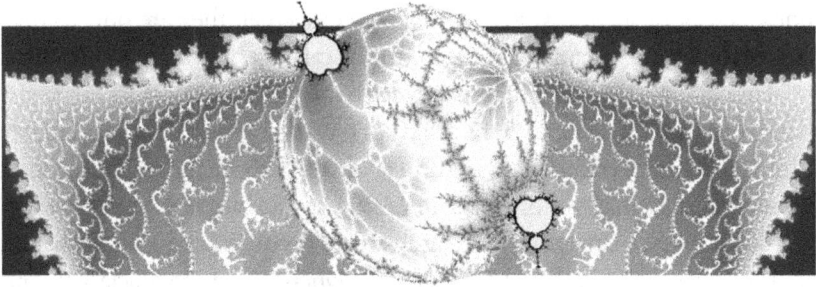

Chapter 2: The Limits of Linear Logic

1. Showdown between the left & right brain

On June 8, 1985, about three months after having that stunning dream of the Double Bubble universe, I faced a showdown between my left and right brain. Shoot-out time. It was late on a Saturday afternoon in Austin, Texas.

For three months now, I'd wondered about that dream, smiled at it, fretted over it. Why would I have such a strange, staggering, overwhelming dream? Me, a humdrum English teacher at UT Austin? But the dream hung fire until that shoot-out Saturday afternoon at Austin's East-West Center, and all because my friend Diana Latham kept insisting that I try the I Ching. Just once? Please?

"Why? It's an oracle, isn't it? Why bother with a superstitious old relic?" But despite all my resistance, Diana would not quit pressing me to try it.

"Come on," she wheedled. "Just once!" Very reluctantly, since she was my friend, I sat down off in a corner and made a lackadaisical, dubious query of the I Ching oracle. I asked a secret question: "Why did I have that God dream?"

The answer dumbfounded me. It was Hexagram 22 with no changing lines. The hexagram of *Grace*. Its title felt poetic or religious—and somehow devastatingly appropriate. Yes, the dream's calm, lucid beauty had blessed me by its grace. It depicted some strange, sublime dynamics that created our universe. It showed me other universes. With a huge love encompassing it all.

But why? Why would a dream reveal to me that wondrous, universal architecture? Such a profound vision of beauty! It came without my wanting it, deserving it, or being prepared for it. Or knowing what to do with it.

Carefully I read Hexagram 22 in the Wilhelm/Baynes translation. Its obscure phrasing had a formal tone. It talked about seeing great beauty that illuminates the heavenly heights. That felt right. It said, "By contemplating the forms existing in the heavens we come to understand time and its changing demands."

That felt right, too. The dream had turned me into a time traveler. It took me back to the origin of the universe, where I saw its beginning. Then I went to a wider scale, and I saw how our universe is just one in a spawn of many universes, all of them diverse. And taken all together, they are only a small part of the Grand Organizing Design that I'll call God for short. So beautiful!

Hexagram 22 declared that such grace is ineffectual if one does not know how to deal with it, so that "all the beauty of form will appear to have been only a brief moment of exaltation." So true. I could not fathom the technicalities of what went on in that wondrous dream. And I had no idea how to deal with it.

But of course, such an apt I Ching answer to my question was sheer chance. Right? Obviously! Logic said so. Right? Yet when I tried the oracle a second time, asking why the dream came to me instead of someone with enough math and physics to comprehend those dynamics creating our universe—if they even worked?—I got the same hexagram answer again. *Grace.*

Moreover, this time the words of that silly old Chinese oracle touched some deep chord in me that logic alone could not reach. I responded with a welcoming internal hosanna of recognition: "Yes, that's it! Sheer grace. Why else would God bother to make love to ordinary, agnostic me? Show me such astonishing, indescribable beauty?"

Hmm. Maybe it came because I was an English teacher? Used to describing things verbally, instead of with math? Because who could ever fathom the math and physics of that wondrous universal architecture? Certainly not me.

But now I was becoming irritated, too. Even angry. Because even if that stupid old I Ching oracle could not possibly work, of course—it was not logical that it could—nevertheless, shouldn't the I Ching go ahead now and tell me how to turn "all the beauty of form" into more than "a brief moment of exaltation" that is then lost?

So I asked another question: "How should I treat this dream?" The answer was Hexagram 14, *Possession in Great Measure.* No changing lines. Hmm, again it felt right. Dammit! I possessed a great dream that no one else could ever see.

Then I realized, "Oh no! If I really do feel this way, certain that this dream is something to treasure, doesn't that mean I need to do something about it? With it? Okay, what?" I did not know.

The I Ching answers felt so out of plumb with my habitual, skeptical, logical stance that by evening, I'd pooh-poohed the I Ching away. How stupidly irrational to dally with that fusty old oracle. Irritated, frustrated, not knowing how to sort out the marvelous treasure I'd seen in the dream, now I locked horns with the very notion that the I Ching oracle might actually work. Hah! Absurd that such an absurd, ridiculous, stupid, bizarre, superstitious, nonsensical (use your own most derisive adjective here) relic might confirm something I was sensing deep inside and wanted to ignore. I could not admit that the oracle might work, even though the notion chewed on me overnight.

2. How can I grasp this?

The next morning I woke up preoccupied. I got up wondering how an ancient, superstition-ridden, outdated oracle could align all three hexagram answers I'd received so precisely with the very private turmoil I carried over that dream? Logic said, "Impossible! Chance! Gullibility!" Yet a simple fact remained: the I Ching had told me apt wisdom, pointed and calm, like a grandparent whispering in my ear. Better than my husband or brother-in-law had managed to do in speculating on my dream, although they'd waxed quite vocal.

Since it bugged me, I took preemptive action. After all, I was modern, savvy, educated beyond superstition. Right? I was a Ph.D. teaching at the University of Texas at Austin. Logic shot down that old I Ching oracle. Right?

I decided, "Why not explore in logical terms whether such an illogical contrivance as the I Ching could possibly work?" My first hurdle was just finding a way to grasp the procedure mentally. It seemed to use a counting-out algorithm. But in school, I'd always avoided math and sciences. I preferred the humanities and the arts…probably because I was better at them. So now I felt handicapped, stranded without any means to debunk that silly old oracle. Or verify it. But why even bother? Maybe to understand my dream better?

Moreover, hard to admit, I sensed a riveting rightness in those three I Ching answers. They buzzed in me a strange shock of recognition, pinged in me a resonance I could not explain since I'd never even seen a hexagram before, having always avoided oracles (too illogical) and math (too logical).

The strange resonance of those hexagrams demanded that I at least give the I Ching some empirical testing. For the next month, I queried the oracle every morning. I asked, "How will today be?" Each evening I compared what had actually happened during my day against the morning's prediction. Decrypting the obscure, ancient text seemed to yield a high positive correlation between each hexagram answer and the actual events that I'd lived out during the day.

But of course, surely I was just creating that effect by being primed for it?

So I reversed the order, and for another month, each night I asked my question after the fact: "How was my day?" Each night the I Ching opened disturbing new insights on my daily hubbub—suggesting that there were underlying patterns and significances that I was only just now recognizing… until I surmised I must be amazingly suggestible and susceptible to that old fraud of an oracle. Even if my friends called me hardheaded, smart, and skeptical. Dubious, secular, and show-me.

Either that…or maybe the I Ching really worked? For 3 months I kept careful records on each forecast or postcast. It was like tuning into my private internal weather report on how to view the climate of my day and deal with it. Despite the verbal thicket of archaic Chinese imagery, its gist made sense. I was adept enough at poetic nuance and scholarly research to realize that this *>&^%!!! old relic probably worked to a degree…somehow! So were those yin/yang answers also a gift of grace, somehow a key to resolving my dream?

But how? I decided to move past my fear of superstition and my lack of math savvy to see if I could discover how and why this oracle worked? If it did. I was still holding onto a reserve of skepticism. But at least now I had a handle on the task: figure out logically how and why it might work. If it did.

I intensified my research. My husband John, older than me, had just retired. Impulsively I quit teaching at the University of Texas and talked John into moving to Zurich with me by September of 1985. John was a jewel. He said, "I've done my career. Now let's do yours!"

My interest was not in a career, though. I just wanted to learn how to unpack that dream's treasure. As I studied psychology at the Jung Institute in Zurich's suburb of Kusnacht, a clue in my quest came from a noted Jungian, Marie-Louise von Franz. In a 1968 essay, *Symbol des Unus Mundus*, she remarked that DNA and I Ching hexagrams have a strikingly similar 64-part structure. After a lecture of hers at the Jung Institute, I asked von Franz about that remark. She merely replied that, yes, it was a wondrous oddity and probably had significance.

I wanted to explore the wondrous oddity myself…but where and how to begin? One can count on numbers, right? Maybe I could take a simple (Ha!) look into DNA's mathematical layout and compare it with the I Ching's hexagram structure. And while I was at it…why not also try to understand the ancient Chinese mindset that had created this extraordinary document that somehow merged abstract math, poetic nuance, and oracular foresight?

Within half a year, I was exploring correlations at night between the genetic code and the I Ching in the downtown Zurich library of the Eidgenossische Technische Hochschule (ETH, the Swiss equivalent of MIT). Why work so hard? It was because by now, I loved the Grand Organizing Design. In acronym, GOD. So maybe I loved God. Why? Such beauty and wonder as I'd seen in that dream deserved attention, even if I must…ugh!…dig into science and math.

Skeptics, especially those of a logical bent—and I used to be one—may try the ancient I Ching oracle once (if they deign to give it even that much time) and say, "Oh, it's mere coincidence that this old Chinese oracle mirrors my inner reality this once, this much. Besides, it's from a language and culture so remote and archaic that you can interpret this stuff any way you want to."

Then they dismiss the I Ching, partly out of scorn, partly out of something deeper that it may cover—a fear, a nameless dread of superstition, of ignorant acquiescence to enslavement in ridiculous folly. At least, that was the case for me.

In that great dream, I saw how the universe began, but for three months, until Diana Latham insisted that I try out the I Ching, I could not begin to fathom how to grasp such grandeur, much less put it into describable terms. But that dream, and then von Franz's words about the genetic code, opened my eyes to ideas so extraordinary that I sought data to explicate them.

Hesitantly I tried to interpret them using our known math and science…I who'd taken no physics or biology in college, no math beyond trig. What a joke! Yikes! I truly did not ask for this task. The only plus from my background was this: by analyzing literature of many kinds, I knew how to do close reading and research. Close reading helped me understand the I Ching. It was like reading goofball poetry that oriented me by its own vatic gyroscope. Research helped with everything else—physics, genetics, chemistry, math.

I gradually realized that the I Ching math figures are not merely binary, as Gottfried Leibniz supposed back in 1703 when he first saw the 64 hexagrams. He reeled, stunned to see that yin-and-yang hexagrams did binary counting, having assumed he'd just invented binary arithmetic himself in Germany.

Rather than being merely binary, I Ching math has a special kind of nonlinearity that I call analinear. Volume 1 showed how its specific range of numbers can merge binary units with analog period-doubling and exponential growth to generate 64 fractal patterns of complementary chaos—or co-chaos. Co-chaos is key to understanding how the I Ching works, how the genetic code works, how their original template, the master code, works to make the universe.

It took me about 5 years to learn how to dive into the depths of universal structure with the waking mind and start sussing out enough secrets to give you this report. Some would call it remote viewing or astral travel. I call it deep-see diving. By now, I've spent many years diving into what Jung called the collective unconscious, what I call the universal mind in the universal body.

Deep-see diving necessarily invokes questions of connection to something greater than oneself. It means juggling physics and philosophy while continually asking, "What are we in the bigger picture? Anything?" The beauty of that great dream drove me to find a way to describe to you…somehow…the astounding majesty of universal origin and what it says about why we are here. More than 35 years later, I am still at it.

3. DNA holds our double spiral of change

While working on this chapter, I had a wordless dream:

I see discrete units marching straight to the solution while exponential variables swirl in an endless, iterating process. Yet they somehow are dancing together.

Waking up, I thought, I suppose that's so. Decisive units drive to sum things up. Exponentials noodle in cycling rhythms, more interested in how things hop around than whether they reach the goal line. If you put the line and circle together, you get a spiral: the supreme symbol of evolution. It epitomizes entrainment plus change; it replicates the old, yet it also goes somewhere new, lifting the monotonously recurring circle of life to higher orders of evolution.

This archetype of the spiral going somewhere is visible in your body's DNA, in dirty water funneling down the drain, in a cyclone, in the Milky Way galaxy's spiral arms. Spirals spin the universe together and knit us into being.

Life needs spirals because it gets tired of replaying the same old tune. Life jazzes itself up by driving somewhere new so that it's no longer just a circle iterating itself, nor is it just a straight line rushing to the goal. Instead, the old theme takes on new direction and scope. It says, "Play it again, Sam, but with a twist. Jazz it up. Take me somewhere new…and make me thrill to it."

Moreover, your DNA employs not just one, but two spirals bonded together in a double helix. Its twisty ladder carries your blueprint. Each rung is a 6-pack of data made by two polarized triplets bonded in the double-entry bookkeeping that stores your genetic code, ensures the protection of its data cargo in your body, and renders survival of the species more failsafe.

In this series, we'll see how co-chaos underlies both the genetic code and the I Ching's hexagram figures. Both variants are templated from the original master code that can be shorthanded by simple yin/yang math.

The modern West found DNA and said, "This code builds organic matter." But the ancient Chinese found the same code long ago and claimed it reveals universal mind in the ever-emergent flow of the Tao. They even handed down a mathematical shorthand for its co-chaos dynamics, plus text from a forgotten rural culture. Its verbal text holds analogies that describe the basic fractal dynamics underlying events. They called it the I Ching, or Book of Changes.

A Taoist code for universal mind. A genetic code for organic matter. Mind and matter…where do they meet? This TOE says they meet far below the quantum scale where matter and energy emerge. They meet at the mobic scale where space and time emerge due to polarized pulsing of the master code.

After more than 35 years now, I can say that our universe is alive, and we are a tiny part of its being. The genetic code shaping your body and mind is a lesser variant templated off the master code that shaped the universal body and mind. The I Ching's ancient math provides a concise shorthand for both the genetic code variant and for the master code that was its template. No wonder the ancients said the Tao cannot be spoken. It must be lived.

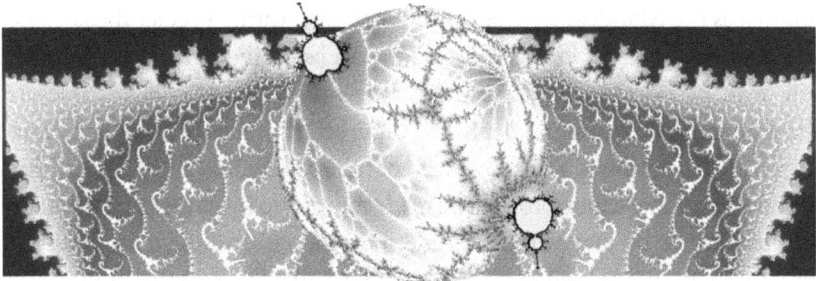

Chapter 3: Looking at Fractal Shapes

1. Mandelbrot's fractal patterns

Benoit Mandelbrot opened a mathematical door onto nature's patterning when he asked a question in *The Fractal Geometry of Nature* that is famous among the chaos cognoscenti: "Why is geometry often described as cold and dry?"

His answer: "One reason lies in its inability to describe the full shape of a cloud, a mountain, a coastline, or a tree. Clouds are not spheres, mountains are not cones, coastlines are not circles, and bark is not smooth, nor does lightning travel in a straight line. More generally, I claim that many patterns of Nature are so irregular and fragmented, that, compared with Euclid—a term used in this work to denote all of standard geometry—Nature exhibits not simply a higher degree but an altogether different level of complexity. The number of distinct scales of length of natural patterns is for all practical purposes infinite."

Mandelbrot turned science's linear approach to nature on its head. He realized that nature reveals to us a dynamical system generating a nonlinear, iterating progression. Our measurements needn't conform the natural world into exact square or round shapes. Instead, we can allow it to relax into the nonlinear shaping and timing that Mandelbrot in 1975 described as *fractal*.

Why did Mandelbrot coin this word *fractal*? He said it indicated that fractals have a "fractional dimension." Thus *fractal* drops a syllable out of *fractional* to convey a strange notion, the idea of fractional dimension. More on that soon.

2. The Mandelbrot Set

The Mandelbrot set is a famous nonlinear system that describes patterning in a fractal geometry of infinite depth, named in honor of Benoit Mandelbrot. Its baroque, ever-changing mandala has been called the most complex, universal number relationship in mathematics. Its awesome beauty comes from shifting numbers that can be graphed as colored patterns on a computer screen.

The Mandelbrot set's appeal rests not just in the vast number symphonies themselves, but also in the visceral thrill of its profound mathematical beauty rendered into shifting colors and shapes around the border of what came to be known as the Mandelbrot beetle, heart, or heart+turtle. I prefer the last one.

Around the dark center's heart+turtle is a lacy fringe that holds Julia sets rife with rich details. Seahorse valleys appear. Dragons. Strange attractors. Imagination enters infinite depths to mine bright jewels of color and form, much as in older times, a treasure hoard of rubies, emeralds, and pearls sparked the imagination. Mandelbrot said he made converts to the study of chaos theory on his lectures by just carrying a slide show along. It revealed an awesome, intricate natural jewelry designed by coloring the shifts of nonlinear numbers.

The Mandelbrot set

Chaos theory attracted many scientists, perhaps because they experienced it not only logically, but also holistically, for the brilliant, shifting colors overwhelm not just sight, but also normal logic. You sense that it somehow holds the flow of nature in it. Mandelbrot made this fractal beauty accessible by using math in a new way. He iterated the equation for the Mandelbrot set's fractal patterning on a computer screen. He colored it in by algorithmic mapping that acts much like a paint-by-numbers kit that can immensely vary the color schemes.

For instance, below are four different colorings of the central heart+turtle in the Mandelbrot set. This shape is often color-coded black, and it looks rather like a turtle pulling a big heart along behind it. I like to think of it as a heart being pulled forward by the ancient turtle of wisdom.

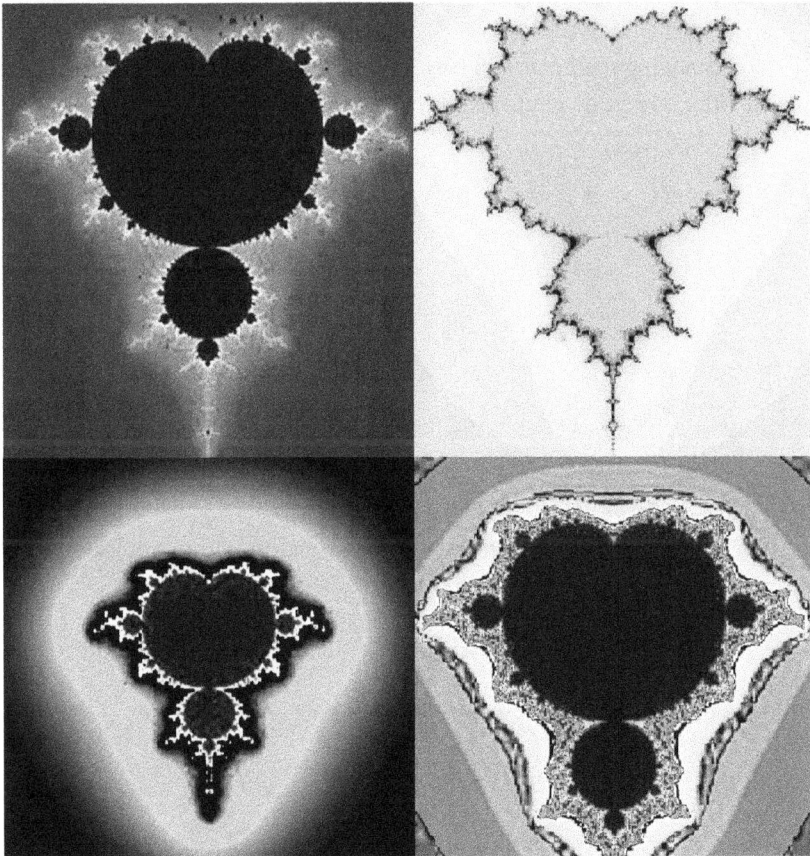

Above are 4 Mandelbrot hearts in the center

Here all four centers are turned upright to emphasize their heart aspect, so they look rather like valentines edged in frothy lace. Usually, however, they are graphed sideways with the turtle-like part set to the left.

3. Here are 4 Julia sets in the fringe

Consider the whirls of tone and shading in the following four Julia sets.

The changing gradients will bestow shifts in the coloring and shading as fractal attractors create continual shifts along their boundaries.

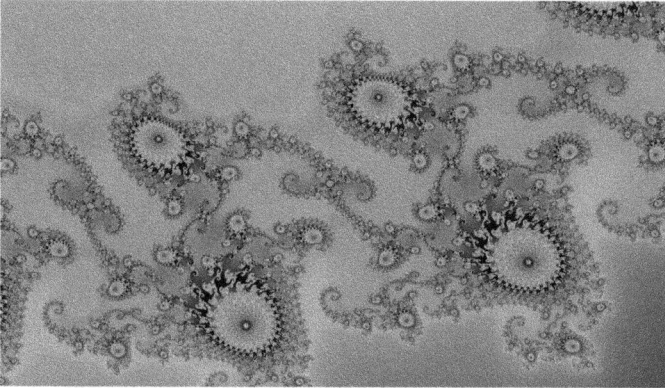

No border ever forms between any two attractors without at least a third attractor insinuating itself. Talk about being betwixt and between!

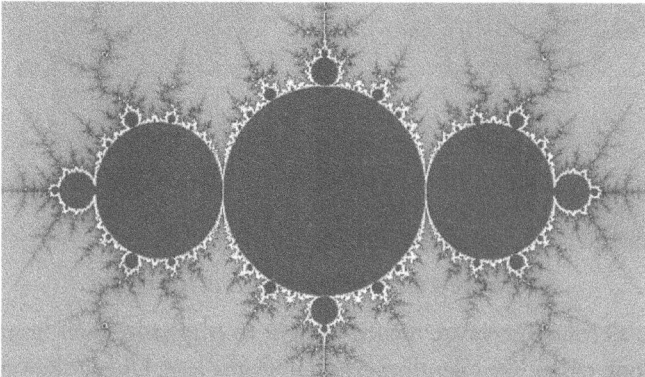

Thus there is no simple *either/or* boundary in these intricate whorls.

Notice the multiple and almost eerie variations on polarized symmetries.

4. Simple equation, complex results...

Simplicity underlies the complex fractal patterning of the Mandelbrot set. Its elaborate beauty is even more surprising when you see the brevity of the short equation that generates it. Here is its short basic equation: $z = z^2 + C$.

To make the Mandelbrot set, you just keep iterating this basic equation on a computer screen using all the real and imaginary numbers that.... Wait! What is an *imaginary number?* Okay, squaring an ordinary *real* number gives you a positive answer, but squaring an *imaginary* number gives you a negative answer. For example, if you square the real numbers of -3 × -3, your answer is +9. However, if you square the imaginary numbers of -3 × -3, your answer is -9.

On your computer screen, real numbers sit on the x-axis of a grid. Imaginary numbers sit on the y-axis. Putting all the x and y numbers together by pairs gives you complex numbers to map points on the grid. In other words, the real and imaginary numbers work together by pairs to create complex numbers that set points on the grid. You can use the equation to iterate an endless sequence of possible complex numbers, producing the Mandelbrot set that is continually evolving on the x-y grid of your computer screen.

The central dark heart+turtle on the grid holds the points of all the *bounded* numbers, i. e., any numbers that do not escalate toward infinity. Instead, those numbers are trapped in repeating loops, or they meander chaotically.

The central dark heart+turtle has a boundary; at that frothy fringe and beyond, the numbers *do* cavort off toward infinity. Those numbers iterate toward infinity in an escalating rush as finer details continually emerge in the shifting structures of gorgeous Julia sets. Their distinctive patterns give the valentine's border its colorful, convoluted lace. This baroque fringe of patterning is infinitely complex, and it holds countless other, ever-tinier hearts.

As the Mandelbrot set's changing patterns unfold in luxuriant beauty on a computer screen, its zoom-lens ability can take a section of detail and keep magnifying it to endless resolves of evolving beauty. According to A. K. Dewdney in *Scientific American*, as this happens, "…a riot of organic-looking tendrils and curlicues sweeps out in whorls and rows. Magnifying a curlicue reveals yet another scene; it is made up of pairs of whorls joined by bridges of filigree.… As the zoom continues, similar-looking objects reappear, but a closer look always reveals differences. Things go on in this way forever, infinitely various and frighteningly lovely."

For the Mandelbrot explorer who tracks the evolving networks of numbers around the central heart+turtle, the shadings of color will step up and down in the shallows and depths of number itself, rather like an airplane view of the ocean's continental shelving along a coastline. The Mandelbrot set iterates its nesting shapes to infinity, so you'll never get to the end of it. You become intrigued by following what happens in the ongoing trip rather than reaching a summary goal…and that is a very yin, receptive way to relate to numbers.

Such beauty holds a visceral impact that is more powerful than mere logic can acknowledge. Scientists and laypeople alike find themselves fascinated by the natural beauty embedded in the Mandelbrot set, seduced by a panorama of majesty whose elegant chaos can only be explored by interacting with it.

The Mandelbrot set itself became a strange attractor to scientists. Why? Because of its intricate, literally endless mathematical beauty. It is self-similar at many scales, it attracts numbered points to certain recurrent behaviors, and its fringe of Julia sets is infinitely detailed. The cumulative effect is mesmerizing!

Those habituated to staying objective and remote may even be disturbed to find themselves relaxing in nature's continually evolving beauty, for Western math and science have traditionally taken a tacitly yang stance that tries to dominate nature by using logic's divide-and-conquer tactics that drive to achieve the climax of a solution.

That goal-driven stance distances the observer from the object, rather than immersing in the ongoing process. Indeed, the linear mindset has often raped and ruined the beauty of nature precisely because it viewed the natural world as something passive, malleable, feminine, yin, conquerable.

Scientists reacted to the wonder of fractals by searching for them in nature, and suddenly they discovered that fractals are embedded everywhere. Now science can apply the theory of chaos patterning across disciplines with a sweep seen in the past only by the blanket edict of some dictator or arbitrary dogma. Yet this broad application bears fruit simply because the theory holds true so widely, so provably, so productively across all of nature's bountiful range.

The impact of this upstart new science is even re-integrating the subdivided disciplines that have fragmented during recent centuries into splinter groups, as topics specialized ever more narrowly to describe still more details over far less scope. But now, patterned chaos research is finding shared ground in areas as widespread as coffee prices, stock prices, road traffic, climate change, celestial mechanics, plasma physics, perceptual psychology, career counseling, and charting the body's living ductwork of blood vessels, airways, and nerves.

Cardiac specialists, psychologists, stockbrokers, and turbulence physicists can all meet at the same conference on chaos patterning. A conference held in Florida recently sponsored papers on chaos patterns in physics, biophysics, physiology, gestalt psychology, chemical systems, mathematics, communication theory, and linguistics. Such unity within diversity was unheard of not so long ago. Here indeed is interdisciplinary reunion!

5. The Silly Centimeter vine

If you're not much into math, how do you get a sense of the Mandelbrot equation iterating its process again and again? Try this:

A friend brings a gift over to your place. It is a pot of soil and a single seed. Your friend says, "I remembered you want a house plant. Okay, just grow a Silly Centimeter vine. This unusual bean here—I bought it at Jack's Weed and Seed—is 1 centimeter long. After you plant it, a vine comes up the next day. Each day the vine grows its current length squared, plus one silly centimeter."

Together you plant the seed in a pot. No growth is showing yet—it's at ground zero. But you know the Silly Centimeter seed is in there since you planted it yourself. You set the pot over in the corner by a window and walk away.

The next day you come back, and the vine is up. How cute. It is 1 centimeter tall. But that is less than the length of your little fingernail! And you did want a nice, large plant for the entrance hallway. Some refreshing greenery. You just hope the plant won't stay too small.

The next day you take a look and find the vine is now 2 centimeters tall. You're pleased by its growth. Good! It seems to be doing all right in this location.

The next day you come back and see that the vine is 5 centimeters tall! A quick grower. Great! Maybe you'll run a string down to it from the ceiling so that it can have something to climb on. You tack up a string.

The next day you come back and discover that the Silly Centimeter vine is suddenly, abruptly 26 centimeters long! Almost a foot high. Great! It's going to look good in that corner, a nice spot of green. You think, "Hmm, I could even train it to grow around the door frame, since it's doing so well."

The next day you come back and—what!—it is 677 centimeters! It is looping in strange patterns over the front door and on the ceiling. And then, the next morning you come downstairs, your Silly Centimeter vine is packing the hallway, taking over the living room! Now it is 458,330 centimeters long! That's over 4 1/2 kilometers, close to three miles!

You decide this vine's seed must have come from Jack's original beanstalk. Looks like you'll have to get someone to come in and destroy this monster plant and haul it off. But whom can you call? A concerned neighbor suggests a tree removal service. You phone Jiffy Tree Removal.

Their truck arrives the very next day, but by then, your Silly Centimeter vine has grown to the astonishing length of 21 kilometers! Over 13 miles long! It twists in tangled clumps throughout your house, pushes against the windows, drapes out the chimney, and has gone out through the front door's mail slot. Neighbors gather in the street, worried at the possibility of multiple home invasions from this monster vine.

It's been only one week since you planted the Silly Centimeter seed, and already its vine has taken over your place and started spreading to the neighbors, creating a huge jungle of green vine that threatens the whole neighborhood!

Of course, the reason this Silly Centimeter vine got so big, so fast, is that it is daily squaring its growth plus adding one silly centimeter, and then it is using that new total as the basis for the next day's growth. Each day the vine cycles the new length into its next iteration of another 24 hours of growth.

Benoit Mandelbrot discovered this nonlinear cycling process when he iterated the short basic Mandelbrot equation in its quadratic recurrence version. It's about as simple as you can get and still throw the discrete lumps of a constant number into the variable number's exponential cycling.

$$z_{n+1} = z_n^2 + C \text{ ...where } C = \text{any complex number}$$

This equation is what energizes our Silly Centimeter vine. At planting, the seed begins at ground zero. No growth on planting day. Overnight it multiplies that nothing by itself, plus it adds in the lumpy constant unit of one silly little centimeter to start a vine tendril shooting up. That first squaring gives nothing, for 0 squared is still just 0. However, since the process also adds in the puny constant of 1, nothing plus 1 = 1. That first iteration completes the first day of growth. The vine is now 1 centimeter tall.

The next day, the process takes that initial growth, the answer of 1, and squares it. Hmm, 1^2 is still just 1...but again, it also adds in the lumpy constant of 1, the silly single centimeter of our vine's constant. Okay, 1 plus 1 gives a total of 2 centimeters. With each new cycle, the process keeps squaring the old answer and then adding 1 more centimeter. Thus it keeps repurposing each old answer to iterate the ongoing process that creates a new answer of another day's growth. In its ballooning cycles, the answer inflates again… again…again. Each iteration says, "Play it again, Sam, and jazz it up in another, even longer variation."

Simple enough, right? The Silly Centimeter vine daily keeps growing exponentially…plus 1. You can watch the rising tide of numbers scream off toward infinity. There is no final product to it, no end sum, only an arbitrary jumping-off point whenever you choose to stop calculating. The process itself drives on forever forward (hypothetically).

Somewhere the Silly Centimeter vine still curls upward, outward, filling up the universe (hypothetically). Each new answer feeds back into the squaring process that keeps its growth going. Each new squared number provides the exponential analog cycling, but the continual addition of that constant 1 also throws in the discrete lump of a unitized number. Voila! You've got an endless dynamic that merges the constant number's chunky unit with the variable number's analog cycling to bring forth a new, third mode that is nonlinear.

6. Many vines grow a fractal pattern

This iterative process can grow more than a single Silly Centimeter vine. That single vine was just one line of growth pushing onward via its iterations. But by planting a whole field of such seeds—say, 10,000—on a computer screen in the proper layout, we can start to grow the whole Mandelbrot set with its dark heart surrounded by a boundary of lacy patterning.

What's the proper layout? The vital trick here is to get some significant variation into the growth directives as we plant all 10,000 seeds on the 2D grid of the computer screen. How to do it? We pick each seed's location by pairing a regular number on the x-axis with an imaginary number on the y-axis. The resulting x-y location on the map marks the spot for that particular seed.

All the complex numbers give their seeds different but coordinated growth instructions, so they can turn and twist, not just be many boring lines shooting at one angle. Instead, each "Simon Sez" seed has its own complex-number command, sort of like "Take four giant steps north and then two baby steps east…." This provides each seed with more than just a simple "grow up" message; it has a "grow up, down, and all around" message that maps its own vine in new, yet coordinated turns. The overall result tells all the vines to grow, yes, but at different rates and in separate yet related directions.

The seeds all start growing as the computer iterates its calculations. Each vine follows its own directive for how fast it will leap toward infinity, or perhaps instead just loll in meandering loops. Vines that get stuck in repetitive loops or take random paths get trapped in the heart+turtle area, usually color-coded black. But seeds that are planted at the heart's boundary and on beyond will zoom up, their vines twisting toward infinity in remarkably stunning patterns!

We get into our computer spaceship, and from high above, we look down at the twisty tangle of dark, looping vines that got trapped in the jungle-like center. From up here, we can see how those dark vines establish the Mandelbrot heart+turtle. Around its boundary, though, other vines rise in escalating, varying, volatile growth rates that weave the colorful fringe of many Julia sets. Their filigree holds blinding beauty in its endless evolving diversity!

In other words, this design on the computer screen did not come from a single seed growing just one line of successive numbers zooming up like one Silly Centimeter vine. No, the flat plane of the computer screen was gridded and seeded with many dots or seeds. They were supplied by complex numbers in the ever-extendable decimal places lying between 0 and 2 on the map.

And what tedious work it would be if we had to iterate all those growth calculations by hand! Gaston Julia did make them by hand in the early 20th century. Fortunately, our computer calculates the iterations far faster than humans could ever manage to do in such a monotonous and dreary chore. Indeed, this kind of equation is called a monotonic process precisely because it just keeps going monotonously on and on...and on.

Now we drop down close to the computer screen to verify that all the ballooning vines are indeed growing out of their spots on the grid. Yes, they all come from dots planted by the complex numbers at specific locations, and due to long associations in a human mind, those dots are actually called *seeds*.

The sprouting vines from those seeds grow in an intricate number weave that can be color-coded to make the Mandelbrot set's colorful patterns emerging on the computer screen. The equation develops its fractal patterning so magnificently because each seed on the screen has its own planting address, plus its own automatically-coordinated growth instructions. That makes it easy to get such orchestrated complexity and keep evolving on the screen.

In the center sits the central heart+turtle, bordered by the elaborate edging lace of numbers frisking around in their elaborate dance toward infinity. If we begin to explore those endless number resolves within the various Julia sets, we eventually discover that countless other tiny hearts are embedded within the fringe, and *all* of them are connected by a well-nigh invisible filament!

In other words, the heart+turtle repeats itself infinitely in the fringe. No matter how long the run of numbers jumps about on the screen, it makes more repetitions of...well, not exactly the same heart, but instead, what mathematics calls a cardioid shape, for who in that field wants to talk about lacy valentines?

7. Boundary shifts

What Mandelbrot basically did was clock each number's escape velocity and compare their rates by location. Watching when and where the designs changed essentially meant that he watched number events shift in their spacing and timing on the screen. James Gleick described the process in *Chaos: Making a New Science:* "…when a geometer iterates an equation instead of solving it, the equation becomes a process instead of a description, dynamic instead of static. When a number goes into the equation, a new number comes out; the new number goes in, and so on, points hopping from place to place…."

Fortunately for us humans, the color spectrum is an effective way to reveal all those continual shifts in numbers. All the points relate to each other and show their connections via colors that provide one big picture. Colors paint the escalating number shifts and turn them into dynamic fractal patterns that are predictable in their general form but not in their specific, shifting contents. Coloring in those boundary shifts is what paints the Mandelbrot set.

But we do not even need to know about numbers to see boundary shifts. We watch boundary shifts happening in actual matter. For instance, when H_2O thaws from ice into water, it shifts over the boundary of 32°F/0°C. That moment of shift-over is quite evident to us visually, even though it occurs way down at the atomic level, where atoms alter their relationships to each other.

Matter at the atomic level is full of built-in polarities. The poles of atoms often sit in random alignment—as in most water—but the poles of individual atoms sometimes line up to exert a unified force—as in magnetized iron.

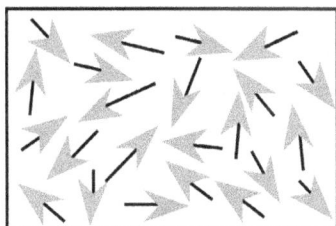

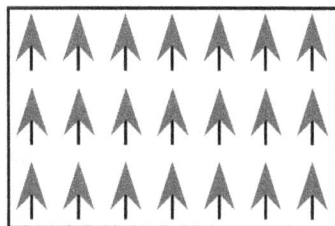

unpolarized water   *polarized iron*

Unpolarized water and polarized iron

Physicists Chen-Ning Yang and Tsung-Dao Lee showed that iron's magnetized polarity at the atomic level has a fractal nature. They revealed variants of the Mandelbrot set by using color mapping according to the magnetic Ising model, named after physicist Ernst Ising. The Ising model allows the mathematical identification of boundary shifts in iron.

Below you see the fractal organization of the Mandelbrot set in magnetic Ising modeling. The same pattern is charted here by two different color mappings.

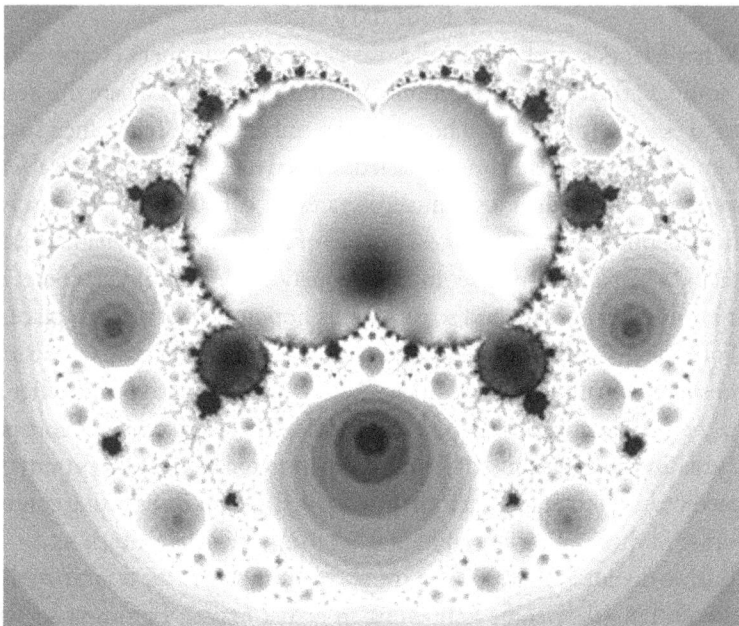

Magnetic model 1 - shows a color-mapped Mandelbrot image

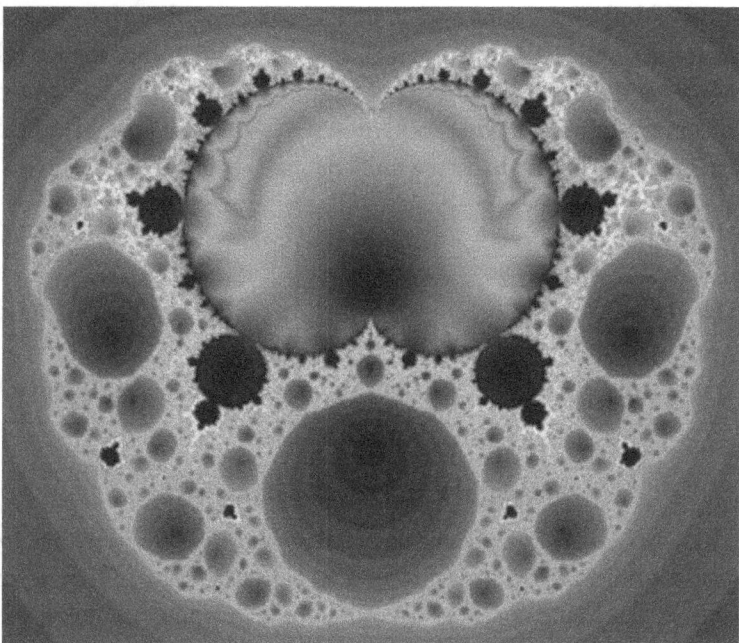

Magnetic model 2 - shows a variant color-mapping of the same image

Both these images come from the same mapping, but they use different color palettes to define it. Each version has a different effect on the eye, but that is merely due to the color choices. The color spectrum's light wavelengths can be quantified by science, yet that result does not provide the qualitative impact of different colors on the senses. These two color images of the Mandelbrot heart in magnetized iron atoms will let us fall away from number's left-brain quantification into nature's visceral impact. Color and pattern affect us qualitatively, which the psyche interprets emotionally. (This effect is not so evident in the black-and-white print edition, but it is striking in the color ebook).

In *The Beauty of Fractals*, Heinz-Otto Peitgen and Peter H. Richter report on investigating the phase boundaries for magnetism in iron. Their account sounds much like two explorers who doubted the legend of Lake Victoria in Africa, lost deep in the wilds of what seem to be random numbers "…bewildering…. This becomes progressively tangled and takes on a random character"…when they abruptly come upon "…a surprise: the well-known Mandelbrot figure appears; its identity with the original Mandelbrot set is astounding."

At that point, two careful German scientists—one a mathematician, the other a physicist—simply declare, "Perhaps we should believe in magic." There is seeming magic in the complex mandala of the Mandelbrot set. But it is not just a flashy gimmick or wordless wonder; it suggests the fractal power that is hidden deep in nature itself. The unseen numbers in nature's fractals have a visceral impact. They impart our full-body-and-brain experience as we live in reality's ever-emergent flux of relationships in color, sound, texture, scent, taste.

8. The web of universal connection

Finding the nonlinear power in nature has begun to heal many rifts that were caused earlier by compartmentalizing life into statistics and logic-chopped data. Starting in the 20th century, various aspects of human knowledge began reunifying as fractal explorers began moving us back toward center. They described a growing web of connectivity that Western-based, linear sciences did not before recognize, a web that was seen earlier only by mystics and artists who experienced it, but could not explain it in logical terms.

This TOE shows the web of universal connection springs from a fractal master code whose ongoing process generated and still maintains our living universe. It provides this universe with a fundamental stability even while allowing its many patterns to undergo major changes. It generates those changes that Chinese sages long ago called the flow of the Tao, a dynamic so vital and significant that the very name *I Ching* translates into English as *Book of Changes*.

Modern chaos theory scientists sound oddly like ancient Chinese scholars who viewed reality as nesting patterns. They used the I Ching to indicate dynamic patterns in the emergent flow of events without being able to predict the exact details. The I Ching oracle works that way; it will show you the general dynamic pattern of an upcoming event, but then you must fill in its unique details according to how you act in the emergent reality of ongoing life. Foresight on those emerging patterns will give your free will a chance to make better choices.

Western science is mostly an elaborate collection of *hows*, but Chinese philosophy was mostly a collection of *whys*. It described life more often in terms of relational, analog qualities rather than linear, unitized quantities. It spoke of reality as dynamic interactions to be experienced personally, subjectively.

Meanwhile, the West has sought to describe nature objectively, maintaining a separate, distancing viewpoint from which to perform the logical analysis of what was often considered a mechanistic, random, meaningless flux of events. Due to the material success of Western science and technology, for many, the notion of ultimate meaning became a silly delusion. Physicist Steven Weinberg showed that mindset in his book, *The First Three Minutes,* when he said: "The more we understand the universe, the more pointless it seems."

The Tao's watercourse way carries a gradient toward meaning, even if you cannot make out the topology of a specific event. In the fractal landscape of reality, the Tao flows toward the sea of meaning. Struggling to swim upstream against its flow is counterproductive and wearisome, but using the current to tap its power by adjusting your course can be enough to revive and reorient you. In following the watercourse way, you discover who you really are, the tasks and joys in life that you are uniquely here to explore. Fulfilling your own unique part in the pattern, with its evolving dynamic and significance, opens up your life to transcendent meaning.

From culture to culture, from age to age, we put our psychic projections at the forming edge of knowledge. At the end of the 19th century, many scientists thought all discovery in physics was just about over. But then the 20th century opened up new scales of territory far above and below the old range of Newtonian mechanics. Due to those boundary shifts of physics, the new territories of relativity physics and quantum mechanics sparked scientific imagination. It inspired the study of black holes, dark matter, dark energy, the riddle of neutrinos, the computer chip, MRI imaging…and on and on it goes.

Because we seek to find and honor a truth in the not-yet-known, the numinous always resides just beyond the limits of the known. Along the unknown's endless, active boundary, we may even glimpse the ever-changing face of something that is divine in the grand organizing design.

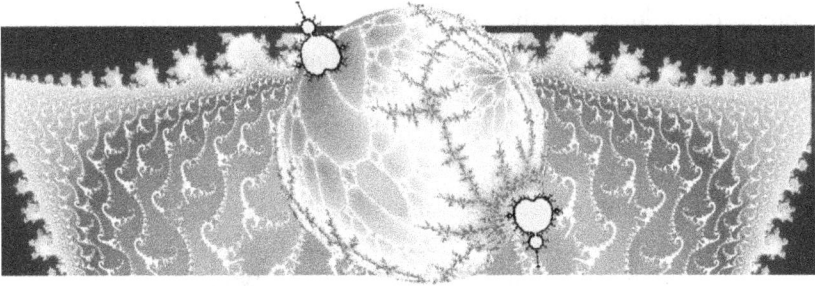

Chapter 4. God Departs

In this even-numbered chapter, I describe teenage events that perhaps put me on track to seeing the Double Bubble universe in a dream, then triggering the subsequent journey of exploration. If not for the watercourse way of random-seeming events that came before that dream, I might have refused to go on this journey. You see, I'm trying to relate the analog resonances instead of just sticking to a dry list of facts so that you realize this series of books is not a random exercise or abstract theory I am pursuing in a think tank.

1. Early days

The family finances were rocky when I was little. My father's small businesses generally failed at the rate of one or two per year, mainly because he was a naive and honest farm boy still trying to figure out how to adapt to the wiles of city life. City meaning Waco, Texas, population around 80,000 at that time, and the world capital of Southern Baptists.

When I was 7, Dad started a dry cleaning shop near Baylor University. Maybe this time, things would click. He tried hard, but the business struggled. As a sideline, he stuck a workbench for doing watch repairs into a corner of the shop.

When not manning the steam press and naphtha fluid extractor, Dad would sit down at his workbench to repair watches and broken jewelry in the slack hours. He'd patiently reset a flywheel on a coed's watch or mend tiny chain links in a broken necklace. Tedious, some would say, but Dad found it soothing, although it involved more time than money. At least he was keeping busy.

As an afterthought, Dad also bought a few tuxedos to rent out. He propped the torso of a headless, legless, armless mannequin into the display window up front. It was outfitted in tux regalia: black jacket, white shirt with stiff-pleated front, red bow tie and cummerbund, gold-toned cufflinks. The coat sleeves hung limp. The black trouser legs were folded and tucked to trail behind.

But that display worked. It attracted glamour-hungry Baylor students who walked past the window daily on their way to and from classes. Dad rented out all seven suits hanging on a rack most weekends, and occasionally he even undressed the dummy in the window to provide an eighth suit.

The big financial outgo was Mother's tuition on her 12-year trek through college to a bachelor's degree. It took her that long because Baylor tuition was so high and she also had to work in the shop most of each day. She studied whenever she wasn't serving customers, altering trouser lengths, sewing buttons back on shirts, shining shoes, counting cufflinks, socks, and bow ties.

Dad thought her tuition was worth the cost. If a life insurance salesman dropped by to sell a policy (and many did right after World War II, when new veterans were roaming out there, seeking a way to re-enter the economy), Dad would just point to Mom and say, "Her education is our insurance policy."

Dad was a "keep your eye on the pie in the sky" kind of guy. But we could not eat that pie for the longest time...not until Mom finally graduated with a teaching certificate. However, she could not find an open teaching position in the Waco school district. Search, worry; search, worry.

Mom finally wangled a job teaching the third grade in tiny Rosebud, a village about 35 miles away. Dad stayed in Waco to run his business alone, and we three moved out to Rosebud—Mom, my brother Norman, age 4, and me, almost 8. I entered the only third-grade class in school. My mother taught it.

That was a strange school year. Mom pretended I was invisible in the class...she said it was to discourage any whiff of favoritism. I mostly avoided talking with my classmates, maybe because Mom meant me to be invisible.

That year was a solitary blur of book-reading and bike-roaming innocence for me, and in many ways, I look back on it as a bucolic idyll. I became quite introverted, maybe to my detriment. Maybe to my advantage, because I read a lot.

Next year, the fourth grade taught by Mrs. Piper was a relief. Finally, I had some friends. But after 2 years in Rosebud, Mom snagged a teaching job in Waco, so we three trooped back to Waco to join Dad again. We lived smack-dab upstairs over his struggling cleaning shop that was gradually turning into a tuxedo rental shop because that part of the business was more successful.

Things worked out pretty well. With more money and less stress, my parents argued less now. The neighborhood boys kept my brother occupied after school. Me? As a 10-year-old, I wandered around the Baylor campus in my spare time. I learned to be invisible there. I could enter its libraries, museums, unoccupied tennis courts, and somehow remain there unchallenged. I began to use the campus as my own private country club. Albeit a very Baptist one.

2. Teenage double whammy: sexual & spiritual awakening

Around age 12, adolescence kicked in. It brought the double whammy of a new interest in sexuality and spirituality, both of them ambushing me at the same time. What a confusing time that was! I began to eye the guys at school in a new way. I also started combing through the values around me to find something that I could actually believe in.

After school, my brother began to stay indoors at his friends' homes, watching grainy black-and-white TV and gaining weight. I studied or read for fun. I stayed home enough to keep the house clean, but I also roamed Baylor's byways.

The material world did not attract me much. Money wasn't that interesting to me, maybe because it wasn't available. Neither was social clout, not in the old-families town of Waco. We didn't belong to the Cotton Palace royalty left over from the huge plantations that once sprawled along that viscous, black clay soil deposited by the Bosque and Brazos Rivers merging just outside of town.

The area's heavy clay soil had once been cotton fields farmed by forced black labor gangs—runaway slaves who'd been recaptured and brought down to Texas in chains, big-shouldered and strong, tough enough to survive the hard work and an overseer's whip. For me, it left a heavy pall on the local mindset.

Nor was I sports-minded. I didn't play any sport well and just watching games as a spectator bored me. And politics? In Texas, it was still mostly a man's game—like football, only rougher.

Nor did I date. I carefully stayed away from dating, along with church. Clearly, I did not know how to handle either one. Not after being invited by a classmate at age 14 to a Sunday night sermon at the Church of Christ, then raped in a car parked outside the church afterward. After that, I still had several male friends, but I made sure they stayed just friends.

3. What do I believe? Anything?

So if I wasn't materialistic, sporty, political, religious, sexy, or a social climber, what was left for me? Hurrah for the inner life! Rosebud had taught me how to explore that. Go within and look for what is true for me.

And what was that? I didn't admire anyone around me enough to adopt their truth automatically. Maybe reading books on religions could help me solve the mystery of how to live, what to believe in? I began to study books on various religions, reflect, then read some more, always still dissatisfied.

During what turned out to be a long trek away from Baptists and Christianity and even God, I slogged on through more religions. They passed by like giant balloons in a parade of puffy good intentions gone droopy over time, flattened by escaping stale air. I wanted something with the living rhythms of nature in it. I did not want directions on how to fit into a Procrustean bed of dogma that involved lopping off my head, feet, and genitals.

Three, four, five, six religions I dismissed. I kept on reading, but increasingly it seemed a washout. Then the seventh religion introduced the way of the Tao, and that resonated with me. Less dogma. More nature. No anthropomorphic God, just ever-rebalancing *yang* and *yin* tweaking events into motion. I wanted to follow the watercourse way. Find the hidden harmony in nature's plan.

It all sounded good to me. Good to imagine that such a plan existed, and even more, the notion of following it. But how could I go about accessing nature's plan? The books didn't say. Reluctantly I put away Taoism. After all, I prided myself on being practical, and let's face it, I did not have a clue on how to get into balance and live in harmony with nature, much less with people.

Nor even with my parents. I remember one night they were working in the shop while we kids did lessons there. Dad had accidentally shrunk the jacket of a Baylor student, courteous Hatsu Hegi. Now Mom was tailoring a new jacket for him as he talked with them about his Japanese name. As they talked, Dad remarked that another of his customers was named Matsu Megi.

I liked the rhyming of those two names, so that night, I put them into a little song. My parents called it a silly waste of time, not a captivating syncopation. Okay, my parents just didn't get me. Well, does anybody's? They were good-hearted, well-meaning folks who walked a constant tightwire, pushed to the limits of their finances, nerves, patience, time. Steely-eyed with judgment, I did not want to emulate them in my adolescent quest for the truth of how to live.

4. I dismiss religion; I renounce God

When I began reading on the eighth religion, Zoroastrianism, I hit a passage that stopped me cold. It said one particular sub-sect of worshipers dug their holy sites below ground to get closer to Satan. They believed they should worship the "Evil One" since God would not hurt them. It was a canny notion, but the idea of currying anxious favor from a devil sickened me, so I quit reading about religions and gave up on the whole God thing.

Maybe religion wasn't for me. None of them. Sure, the adherents meant well, but they'd done terrible acts in the name of God. And under the weight of centuries of dogma, bureaucracy, ritual, custom, how could any religion retain much contact with the ever-manifesting edge of truth? My opinion? Religion turned into dogma that people supported more than it supported them.

One afternoon, I stood at our house's back screen door and looked out through its wire mesh at the blind, blue sky. I was pondering a not very bright, polio-crippled girl in my 8th-grade class. She dragged one leg behind in a metal brace. Mute, mousy, mopey Marilyn, ignored by the other students.

Marilyn doubted her own every act in math class, even down to writing the date and her name on a daily worksheet. Often the teacher had us exchange papers across the aisles to student-check the answers. I saw that Marilyn always dotted the *i* twice in her first name and crossed the *t* twice in her last name.

One day I asked her why she did that. She answered, "Just in case."

Why would a God do that to her? Make Marilyn doubt her own simplest act? Thinking about it, I shook my fist at the sky out beyond the screen door. I cried out loud to the oblivious sky—foolishly, I admit—"Even if you did exist, God, I would not worship you! I won't worship a God who makes this world so sad! There's too much indifference from you, too much pain here. Such suffering! Such sorrow! Why, I could do it better myself!"

Talk about adolescent hubris! Talk about asking for trouble! Not only did I judge religion and find it wanting, I even judged and found God inadequate. As if I had the capacity to measure that! But by then, I'd developed a lot of orneriness from being around Southern Baptists. Courage, I called it.

"Face it," I said, "I can't follow a religion. But I can still at least try to be ethical and good and caring. Right? I'll just find a good system of values to live by." Why not? After all, I was 15 by now, able to make my own decisions about what is right and wrong. It would be easy, right?

5. I turn to philosophy

Turning away from religion, I looked at the secular world to find answers on integrity, justice, and truth. Gazing at the society around me, I could not find satisfactory answers there, either. Oh, such sorrow, fear, loss, conflict. Such raucous greed and narrow-eyed competition. And oh, the pointless pain!

Seeking a good system of values, I began to read philosophy, but with a growing sense of desperation. I started at the ancient Greeks and plowed right on up through the centuries' historic ranks. Western ranks, that is. The Waco Public Library did not have any Eastern philosophy on its shelves, or maybe I just didn't find it. Anyhow, one book I read claimed that Eastern philosophy was mostly embedded in its various religions, so I'd already dismissed them.

By now, I was a fast and weary reader, scanning through centuries of turgid philosophical prose, looking for a code to live by. My favorites were some of the early Greeks with their schools of specialized, if peculiar, clarity. Sophism. Epicureanism. Cynicism. Stoicism. Wow, they paraded by like different breeds of dogs, one so totally of its own conformation and temperament.

Doggedly I churned through a millennium and a half to emerge smiling, lounging for a moment on Roger Bacon's oddly spiritual pragmatism. And then I found Spinoza and loved him for his mystic heart. But Hegel flabbergasted me, confounded me. What a chill, skilled obscurantist! And Schopenhauer? He made me cry at some invisible Teutonic Wailing Wall of what never was.

At age 16, I came upon Kierkegaard. Oof! Gloom and doom! Walking the razor's edge of free will over obsidian shards of despair glittering in a hell that constantly lurks just below your attention. Not real hell, mind you. The kind you carry inside, an agonized isolation from meaning itself.

Hey, why did life have to be so fraught? I was a serious kid, but maybe not that serious. Still, Kierkegaard impressed me with his idea that religion depends on revelation, if not faith. Sans a revelation, no wonder I couldn't swallow it.

I trouped on through the European existentialists—Nietzsche, Heidegger, Camus, Sartre—and they just wore me down with their mental masturbation of angst, boredom, alienation, life's absurdity, the nothingness of existence... until I finally just caved in from it all. Brainwashed. Depressed, I quit reading.

Remember, this was Continental existentialism I'd been immersed in, back before Jack Kerouac went *On the Road* to party up the existential movement, trip it out, and cast a jokey tone over the whole Beat Generation.

At 17, I contracted existential despair like a disease. Symptoms? A bleak Euro-trashed view of life and death. No God. Isolation in an uncaring universe, each separate intelligence locked in its own flabby prison of flesh that inevitably must partition us all off, one from another, so only by clutching at lonely honor can we ennoble our meaningless lives. Here was alienation with a vengeance!

Yep, in the summer before my senior year of high school, when a teenager is due to commit some adolescent folly of drugs, drunkenness, crime, or sex, I did something much more quiet and damning. I kicked out God.

At that very juncture, on one hot August night, God graciously appeared in a dream to say goodbye. But this was no antiquated divinity left over from some religion I'd read off the library bookshelves. Nor was it the frail wisp of an anti-God cobbled up from my newly spartan, make-do philosophy.

No, this was instead a non-denominational, universal God, who just happened to show up as a larger-than-life, handsome, flaming-haired young male...and I do mean with hair of real flames! God was athletically-built, calmly handsome, and stood over 7 feet tall. He wore nothing...nothing but a blurry glow around the genitals. Talk about impressive! Here is that dream:

6. I dream of God as a teen idol

I am sitting in an auditorium. The dim light in here has an inviting, golden glow. The walls are draped in dark red velvet. Somehow light doesn't quite reach into the corners. It makes the room look rounded and cozy.

We sit in rows, waiting. When a number is called, somebody stands and walks up the aisle and outside. Eventually, my number is called.

Nearly 30 years later, while at the Jung Institute in Zurich, I finally realized why the light didn't reach into that room's corners. It was because there were no corners. That red velvet auditorium was the collective womb where we all sit waiting to be born—you, me, all of us. It is the rounded womb of time. Looking back from this distance, I suppose now that the number called out was my birth date. That's why I got up. Anyway, in the dream...

I stand and walk up the aisle. I pass through the double-door exit to the outside. I stand looking into the black void beyond. How beautiful! I can see the ball of Earth way out there in black space. On the globe, the greens and browns of continents show in mottled relief. Oceans scintillate in blue sheets below the swirls of white clouds. Flashing oceans and moving clouds...it all looks so alive!

This dream occurred back in 1957, 15 years before Apollo 17's moon trip gave all of us down here on Earth a Blue Marble view of this planet.

In the dream...suddenly I'm right down on Earth. I'm living in a village. I even know its population size—about 10,000—and I also know that many such villages dot the planet. I also understand that each community is small on purpose. We mostly favor living in settlements of around 10,000. Why? Because that's big enough to provide culture and commerce, but still small enough to keep us in close harmony with each other and nature. Plus, there are traveling bands of artists, craftspeople, and traders who move constantly from one settlement to another.

How do I know this? Hey, in dreams you just know things.

I also know that this rural-urban blend allows us to enjoy nature without fighting it. We work enough to hone our ingenuity. We play enough to savor our lives. We meet and talk and get to know each other in daily events, so we become informally accountable to one another. We are both grounded and spiritual, both deliberate and spontaneous. We are whole inside, and it reflects in our communities.

Again, how do I know this? Because it is a dream, and I just do.

All of this is wonderful, but the best part is that God walks among us. Walks around in this village, and in all the villages, and it's perfectly normal. We live in easy relationship with God, with each other, with animals, with nature. We love this immense, world-wide, varied garden where we live. God is our friend.

Then God comes up to me. He touches my arm gently, my left forearm. His touch is so sweet, but the words that he speaks—oh, how they burn! God's touch has the sensation of balm even as his words scald my being like molten fire.

He says, "I am going now. But remember, I will come back."

Then God is gone. I am dazed with grief. Sunken in loss.

7. God departs...and good riddance!

That morning I awoke right before dawn. I felt stunned, confused, bereft, and yet...I also marveled at how God showed up in a dream and overwhelmed me by the power of his presence…when just the day before, in the waking world, I'd intentionally, formally kicked out the notion of God as a silly fantasy.

Yet here in the dream, God came up and touched my 17-year-old left arm… oh, so gently. How tender, that gesture! Now I suppose his touch was maybe to anchor into my flesh a faint memory of the promise that God would return.

Once I woke up and got up from bed, my ego came online in that pre-dawn gloom and reoriented me back into my bleak, blasé, adolescent skin and attitude. What I felt was not sorrow at God's leaving, but instead, scorn and derision at myself for dreaming up such tripe! How ridiculous! What a flame-haired hunk of a God I'd conjured up! A silly teenage heart-throb!

I looked around my familiar bedroom with its tweed-toned wallpaper that I'd picked out, pasted, and rolled up myself. I pooh-poohed the ridiculous God who looked like a supercharged teen idol. And oh, his portentous exit line: "But remember, I will come back." Was it lifted from Christianity? A Jesus-clone gone pyrotechnic in a crown of flames? Was this some resurrection schtick?

But oh, that lovely village in the dream—to me, it felt suspiciously like my doubtless too-idyllic memories of running across the connecting lawns of friendly Rosebud so long ago, back when I was just a mere child—what?—8 or 9? How juvenile. Why would I have such a stupid dream?

"Oh," I crowed. "Now I get it!" According to Nietzsche, Sartre, and Camus, that farce of a dream must signify some fervid teenage anxiety, some puerile fear that I wasn't even aware of. "Of course! I'm just about to start my last year of high school. Close to leaving home. This dream means I'm afraid to leave the parental security blanket. I dreamed up a God to step up and take care of me when I leave." It was a wish-fulfillment dream! (Even though at least when awake, I yearned to leave home.)

"How cowardly of me! Nietzsche, Sartre, and Camus are right. This insecurity is the weakness that every religion exploits, that Freud pinpoints, that Sartre decries." Obviously, the solution was to get rid of God! Unnecessary crutch! Stupid old fairy tale! Out, damned frailty! Henceforth, I would demand logic, reason, tangible proof! God was dead to me now, and good riddance.

At age 17, I did not recognize the significance of that dream. It wasn't to ease a separation from my parents. It was to ease a separation from God, who walked up, gave me a message, and left. But it took 27 long years for the message to sink in…until finally at age 43, I had a dream where God came back. That dream of God returning is in Volume 1, *Double Bubble Universe*.

Those two dreams are no doubt what triggered this series of books.

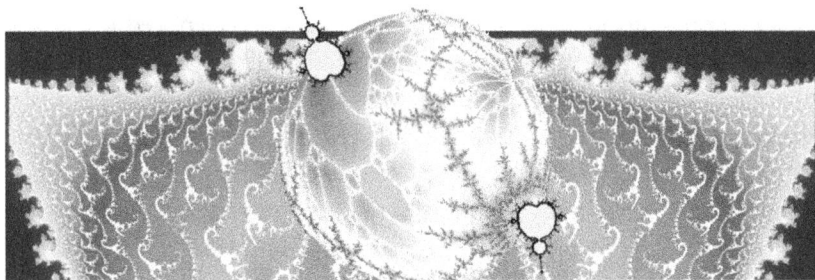

Chapter 5: Measuring Fractal Shapes

1. The fractal snowflake

A fractal is a triumph of utility. It iterates a simple process again and again with an ongoing feedback loop that repeats the basic form with continual variations. Those variants are self-similar across different scales, iterating in space, time, matter, or energy...or in all of them working together.

Helge von Koch described an interesting fractal pattern in 1904. To make it, he took a triangle and "grew" a new, smaller triangle onto each side. He could keep growing new, tinier triangles on each previous version's outer edges, iterating the process again and again. The next graphic needs to sequence through just three stages to reveal the fractal Koch snowflake.

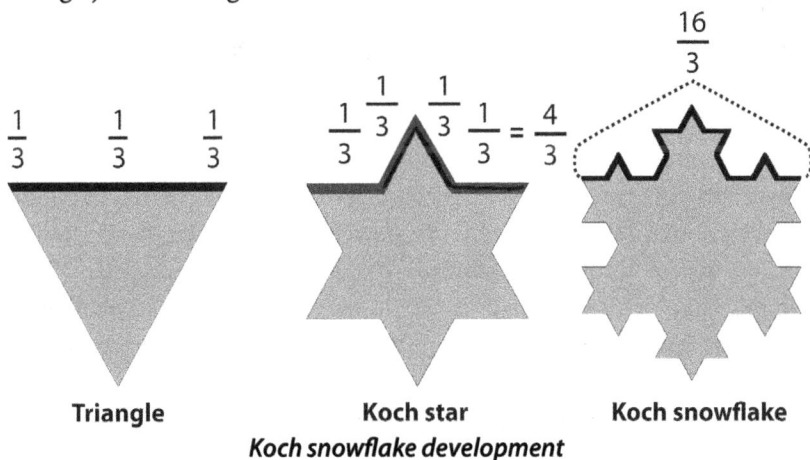

| Triangle | Koch star | Koch snowflake |

Koch snowflake development

Here's how to make it: draw an equilateral triangle and divide each side into thirds, as ⅓, ⅓, ⅓. Thus your triangle has three sides, and each side is ⅔ long. Now add a new, scaled-down triangle on the middle ⅓ of each side's boundary to get a Star of David shape. Your original ⅔ sides have been replaced by three sides that are each ⅘ long. So each side is now ⅓ longer. You've altered your original triangle's shape so its boundary line can make 60° turns to both left and right. Thus this new shape's perimeter is longer and more jagged.

On the next iteration, those three sides of ⅓ length get replaced again using the same format. The result is three new sides with a still-longer perimeter that's even more jagged. Each new side is now $^{16}\!/_{9}$ long, and the original triangular shape has become a serrated disk that looks rather like an Asian throwing star.

At each new iteration, each side's total length will get more jagged and again increase in length by 1/3. In other words, at step n (any number you plug in while iterating the process), its new length on each side will be $(\frac{4}{3})^n$ of the previous side's length. This exponential growth happens at each new stage of development, and it is typical of fractal iteration in a nonlinear process.

This Koch snowflake holds in rough and tiny microcosm the promise of mathematical infinity, for if you keep on iterating the process forever (Hah! May you live so long!), you can piggyback new, smaller triangles onto that original shape nigh unto infinity, at least in mathematical theory.

Each new iteration puts more tiny kinks into the snowflake's perimeter, so if the process really did go to infinity in time and space, its perimeter would have infinite spatial length. Yet this Koch snowflake could still fit into your hand! How can that be? I mean, practically speaking? Cosmologist Max Tegmark summed it up this way: "Infinity is a beautiful concept...and it's ruining physics."

As the prickly little Koch snowflake keeps on iterating, it becomes full of tiny protrusions sitting on bigger protrusions, again and again, all around the perimeter. It's like an island of bigger peninsulas protruding with smaller peninsulas *ad infinitum*. On this hypothetical island's jagged "coastline," the water level sets an arbitrary slice across whatever comes down or up to meet it—cliffs, boulders, sand. Wherever the water meets the land, the waterline in effect draws a jagged silhouette of the island's shape, showing its fractal outline.

A version of this even happens in ordinary life. If a satellite photographs all 16,000 miles of Australia's coastline from 22,236 miles up, its rugged outline is reminiscent of what you'd find walking along the water's edge to photograph just 1.6 miles of it. The smaller version carries a rough fractal profile that's similar to the larger version photographed by the satellite, just on a different scale. This self-similar scaling trait of fractals is locked into nature itself.

2. Spiritual views of the Koch snowflake

Long before Koch described his fractal snowflake mathematically, it was used symbolically and decoratively in various religions and countries. Older traditions saw the 6-point star as a union between male and female, or as the Star of David. The next image shows two triangles on the left. In the center, they unite into a 6-point star. On the right, a design looking rather like a lacy doily holds a proliferation of so many stars that it suggests generative creation.

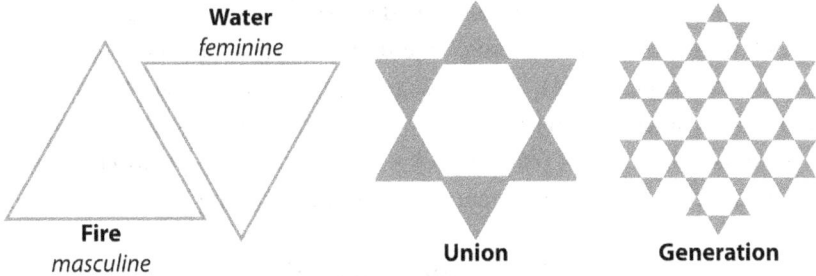

Koch snowflake propagation

The multiple-stars doily, above right, has a variant that's shown below in the center of a Nepalese yantra. A yantra is a meditation on the rhythmic timing of energy, whereas a mandala is a meditation on the harmonious spacing of matter. From the holy order in its sacred central *temenos*, 4 T-gates open out into the ordinary world beyond. In that sacred center, 6 Koch snowflakes form a protective perimeter around a 7th snowflake that signifies the centering spot of divine union. The stars are encircled by a fringe of 16 petals—8 dark and 8 light.

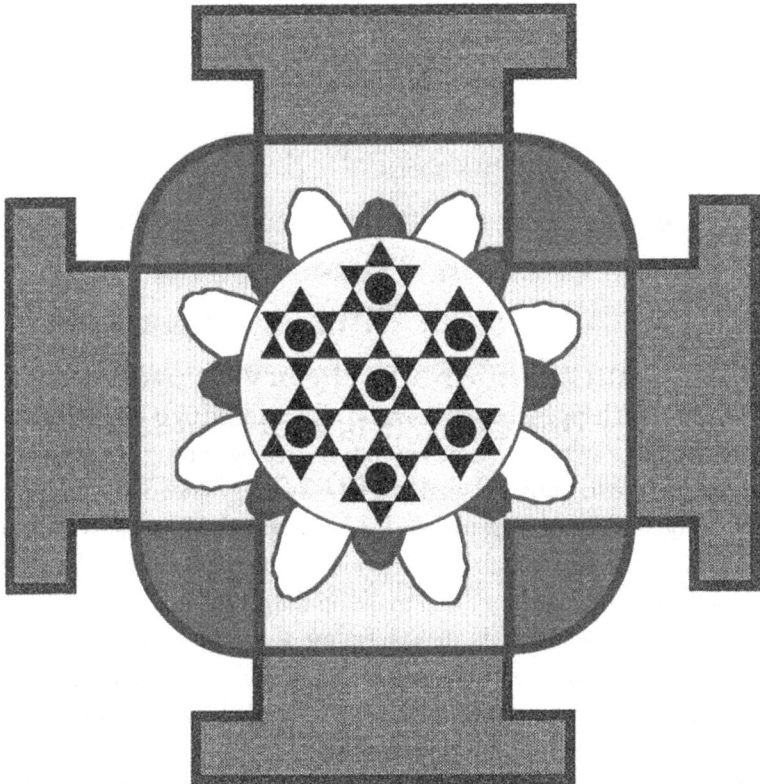

A yantra from Nepal

This yantra references the core paradigm of dark yin and light yang that the I Ching uses in pair-bonding 8 × 8 trigrams to generate the 64 polarized 6-packs called hexagrams. The paradigm's genetic variant pair-bonds 8 × 8 molecular triplets into 64 polarized 6-packs of DNA code to generate us. The master code variant pair-bonds 8 × 8 griplets into 64 polarized 6-packs of dimensional bonds to generate space and time latticing.

3. The square snowflake

The Koch snowflake is based on a triangle, but it also has a right-angle version based on a square. The square snowflake jogs at 90° angles to morph its shape. On the square below left, each side is ¼ long. In the center, each side becomes ⅞ long. Its awkward juggernaut shape zigzags in and out in a regular way to double each side's length, yet without increasing the total area.

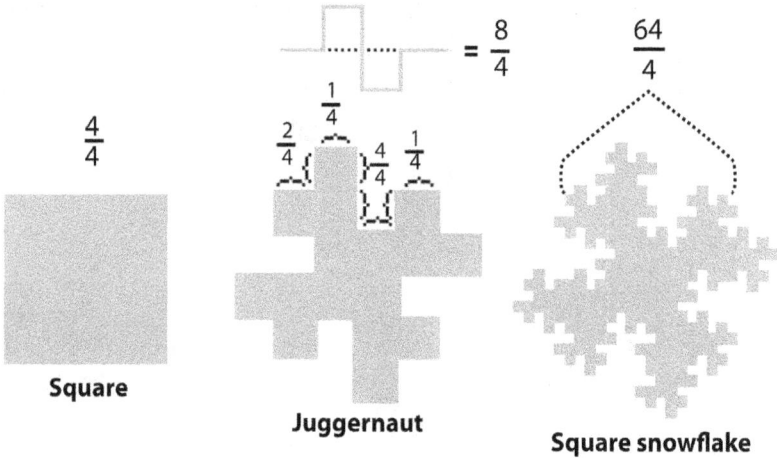

Square

Juggernaut

Square snowflake

Square snowflake development

What? Yes! It happens because each bit of area added by pushing out a new square also takes away that same amount of area by cutting out a square of equal size next to it. Thus, although the four zig-zaggy sides of the juggernaut shape in the middle are now each ⅞ long, the area inside remains the same. You can verify this for yourself by doing box-counting, i.e., bothering to count all the individual square boxes that make up the central image.

On the right, the snowflake shape turns more graceful as each side becomes $^{64}\!/_4$ long. Each side is now 16 times as long as it was initially, yet still with no increased area in its fractal shape. This exponential change keeps happening at each new iteration, which is typical of fractals.

The graphic shows only three stages of development, but there could be many more, again to infinity…at least theoretically. The process could iterate forever to become a truly bizarre, tiny shard that is toothed with infinite points on an infinitely long border. You'd wind up with a square snowflake whose raspy edges give it an outer boundary of infinite length. Yet it encompasses an area so tiny that you, like William Blake examining his grain of sand, could hold infinity nestled in the palm of your hand.

4. Effective dimensionality

Perhaps the strangest thing about a fractal is the idea that it can have fractional dimensionality. Fathoming that involves thinking about the *effective* dimensionality of something. To get this notion, imagine that you live in Auckland, New Zealand, and you want to fly to Seville, Spain.

It will not do you much good to demand the truly shortest route between Auckland and Seville. Nobody will get a drilling machine and bore a tunnel through the solid globe to guarantee you the shortest route between these two cities that happen to be almost exactly antipodal.

No, you'll have to take an airplane flight around the curve of the earth. To make a long-distance flight, a pilot can locate any spot on the globe using just two coordinates: longitude (length) and latitude (width).

However, your flight path from Auckland to Seville is much like following a long, kinked string because your plane has to make a stopover in Bombay. It flies in effective dimensionality—a long, kinked length. So how does a length of kinked string acquire more dimensionality? Where does the 1D line of string kink into? Another dimension? Well, partly. To a mathematician, the flight path *starts* into another dimension. But it gets only partway there, fractionally there…rather like the kinked line along one side of a square snowflake.

But unlike a flight path, a Koch or square snowflake iterates kinks along its boundary in a regular, patterned fashion as it sits on a piece of paper or a computer screen. The regular pattern of that iterative process not only pushes toward a second dimension, but it also can be described fractally. That boundary is like a line of string that kinks regularly and more frequently at each new iteration. Fractional dimensionality is basically a number that tells you how completely the fractal shape is filling space at each new iteration.

"Wait a minute!" you may protest. "That sounds like a math trick. Common sense says we only have whole dimensions! Space has three dimensions: *width*, *height*, and *depth*. Okay, throw in Einstein's oddball fourth dimension of time. See, it still doesn't add up to a fractional total of dimensions."

But are you counting right? This TOE says the upper bubble has ½D time. The lower bubble has ½D space. Each is a vector in the tensor network of a single, yet ubiquitous, polarized dimension that 8-loops across both bubbles.

5. Find the fractal dimensionality

The Koch and square snowflakes both have fractional dimensionality that pushes from 1D space into 2D space using fractional space-filling capacity. How do you figure out each snowflake's fractional dimension? It's easy.

Step 1
Subdivide a side by the number of sides. Find its fraction.

$$\frac{1}{3} \quad \frac{1}{3} \quad \frac{1}{3} = \frac{3}{3}$$

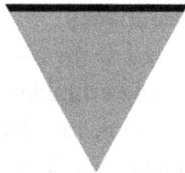

$$\frac{1}{4} \quad \frac{1}{4} \quad \frac{1}{4} \quad \frac{1}{4} = \frac{4}{4}$$

Koch snowflake   **Square snowflake**

Step 1:Subdivide one side of each snowflake

Above on the left, to start the Koch snowflake iterating, first count the triangle's number of sides: **3**. That black line atop it could fit around its sides 3 times. Then to start the square snowflake on the right, count the square's number of sides: **4**...and the black line atop it could fit around its sides 4 times.

Next, fractionate the black line atop each shape according to that shape's number of sides. The denominator will be the number of sides on its shape. The numerator will be 1. The triangle's side equals ⅓, but the square's side equals ¼.

Step 2
Iterate the fractal shape. Find its new fraction.

$$\frac{1}{3} \quad \frac{1}{3} \quad \frac{1}{3} \quad \frac{1}{3} = \frac{4}{3}$$

$$\frac{2}{4} \quad \frac{1}{4} \quad \frac{4}{4} \quad \frac{1}{4} = \frac{8}{4}$$

Koch snowflake   **Square snowflake**

Step 2: Iterate each snowflake's shape

The Koch triangle iterates by projecting a new, tinier triangle on each side, so each side becomes ⁴⁄₃ long. But the square version iterates by projecting a small square and also cutting a small square into each side, so each side becomes ⁸⁄₄ long. Fortunately, as the iterations of both snowflakes continue to jog upward to new levels of complexity, they stay self-similar. This means the first two steps are enough to let you figure out each snowflake's fractal dimension. Then you'll know its consistent rate of space-filling capacity.

Step 3 Make a fraction out of a snowflake's two steps.

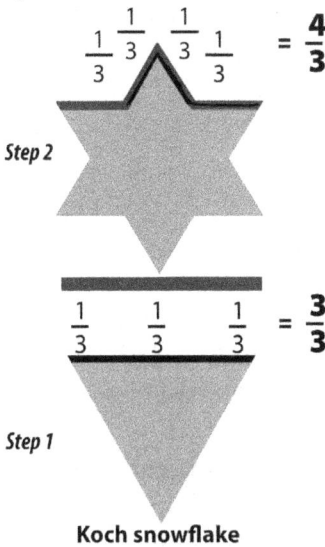

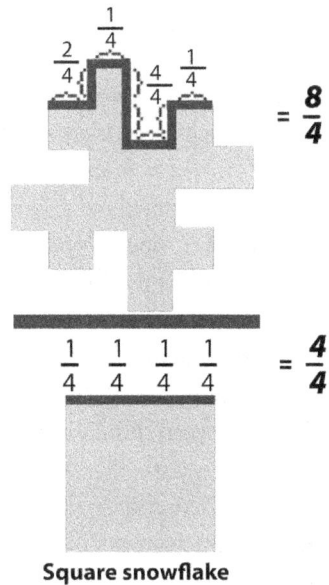

$$\frac{1}{3}\ \frac{\frac{1}{3}}{}\ \frac{\frac{1}{3}}{}\ \frac{1}{3} = \frac{4}{3}$$

Step 2

$$\frac{\frac{2}{4}}{}\ \frac{\frac{1}{4}}{}\ \frac{\frac{4}{4}}{}\ \frac{1}{4} = \frac{8}{4}$$

$$\frac{1}{3}\ \frac{1}{3}\ \frac{1}{3} = \frac{3}{3}$$

$$\frac{1}{4}\ \frac{1}{4}\ \frac{1}{4}\ \frac{1}{4} = \frac{4}{4}$$

Step 1

Koch snowflake **Square snowflake**

Step 3: Set each snowflake's two steps into a fraction for evaluation

For each snowflake above, study its change from Step 1 compared to Step 2. To do that, set both steps into a ratio or fraction. Make Step 1 the denominator on the bottom, and Step 2 the numerator on top, with a fraction line between them. Now evaluate how much each snowflake changed by going from Step 1 to Step 2.

Step 4 Make a new fraction out of each snowflake's two numerators.

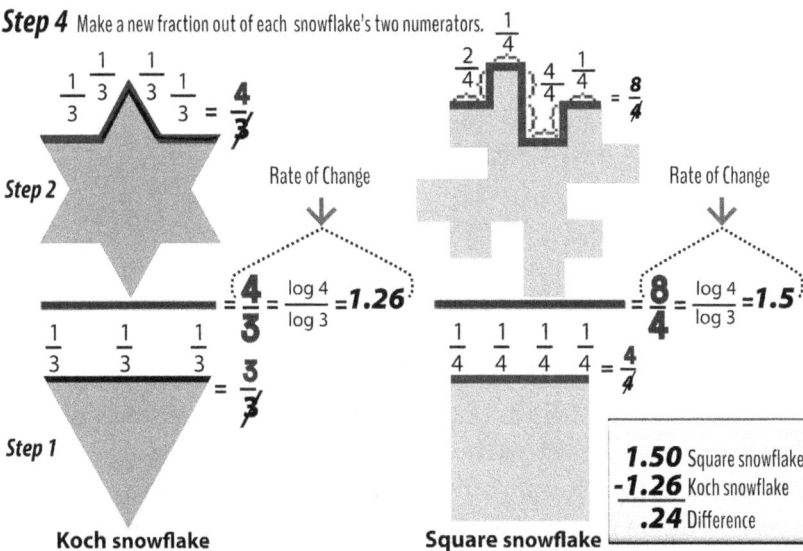

$$\frac{1}{3}\ \frac{\frac{1}{3}}{}\ \frac{\frac{1}{3}}{}\ \frac{1}{3} = \frac{4}{3}$$

Step 2

Rate of Change

$$= \frac{4}{3} = \frac{\log 4}{\log 3} = 1.26$$

$$= \frac{8}{4} = \frac{\log 4}{\log 3} = 1.5$$

Rate of Change

$$\frac{1}{3}\ \frac{1}{3}\ \frac{1}{3} = \frac{3}{3}$$

$$\frac{1}{4}\ \frac{1}{4}\ \frac{1}{4}\ \frac{1}{4} = \frac{4}{4}$$

Step 1

Koch snowflake **Square snowflake**

1.50 Square snowflake
-1.26 Koch snowflake
.24 Difference

Step 4: Find the difference in their rates of change

Step 4 compares their rates of change and shows which snowflake pushes further into 2D. To find this out, take each snowflake's two fractions from Step 3 and use only their numerators to develop a new single fraction for each. This will show each snowflake's rate of change, i.e., its space-filling capacity or push tendency as its boundary pushes from 1D toward 2D in each new iteration.

The Koch snowflake's rate of change is ⅓. To refine that answer further, you can type into Google *log 4/log 3* to get the answer of 1.26. How to interpret it?…1.26 means the Koch snowflake's "push tendency" into 2D is .26 more than 1 dimension (a line), but it is also .74 less than 2 dimensions (a plane).

But the square snowflake's rate of change is bigger. It is ¾…or log 8/log 4…or 1.5…which means the square snowflake's "push tendency" into 2D is halfway between dimensions 1 and 2, for 1.5 is .5 more than 1 dimension (a line), but it is also .5 less than 2 dimensions (a plane).

Finally, here comes the last comparison as we reach this examination's end goal: how much greater is the square snowflake's push tendency into the second dimension, compared to that of the Koch snowflake? To find out, we subtract .26 from .5, and we discover that the square snowflake's push tendency is .24 larger than the Koch snowflake's push tendency.

This difference occurs because the triangle had only three 60° angles in Step 1, while the square had four 90° angles in Step 1. The square's greater number of sides and bigger angles allowed its fractal to push farther into the next dimension at each iteration than the Koch fractal could manage to do.

6. Patterns as waves in matter & mind

How do patterns relate? The pattern below could be a Moorish tile…

Tile 0—Basic Wave Pattern

…but imagine that within its black lines, the white spaces are transparent. If this transparent design is replicated in rows, and then the rows are overlaid so they can slide along atop one another, the resulting images will vary, depending upon how the overlaid patterns of lines move in and out of phase.

Following are three instances of sliding the overlaid tiles along a row. Note the 3 tiny numbers atop the row; they identify the 3 "tiles" as separate images.

A physicist might view these line interactions as patterns of constructive or destructive interference. But most folks would probably just notice their own subjective impressions of the dots and lines.

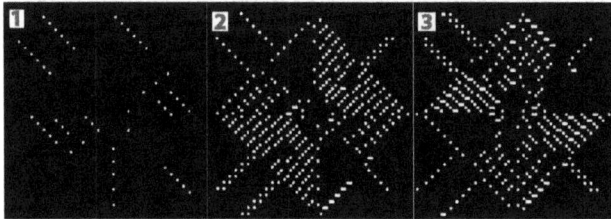

Tiles 1 through 3

Personally, as I scan across Tiles 1 through 3, not much within the dots and lines really pings any image associations. In Tile 1, the amount of interference is so obscuring that I cannot discern or imagine any actual design in it. Tiles 2 and 3 offer somewhat more in the way of possibilities to my eye, yet I still get no strong pattern response by looking at them. How about you?

However, the next four tiles sliding along the row show patterns that seem more visually associative…in my admittedly subjective view. They trigger some networks of association that call up fanciful images, at least for me. A psychologist might liken them to images in a Rorschach test.

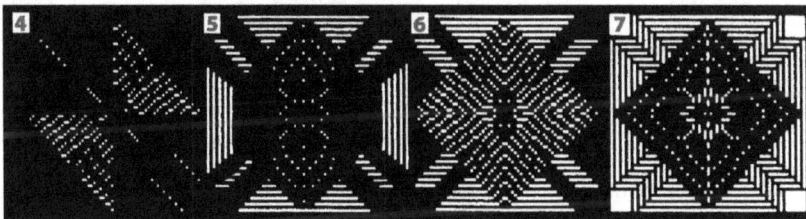

Tiles 4 thru 7

In Tile 4, I perceive a butterfly. In Tile 5, I see a spider. In Tile 6, I spot a small bee on honeycomb. And the final Tile 7, to my mind, improves upon the original tile design by making it more dramatic and interesting. My own reactions to all seven of these tiles suggest to me that their wave interactions vary not only quantitatively in the designs, but also foster varying qualitative effects.

Each interference pattern in the overlaid tiles evokes a different response in me, much as my relationships with different people evoke different interactions between us, both quantitatively and qualitatively. I am trying to suggest here via imagery that when several waves of energy relate constructively, they amplify each other's power. When they interfere destructively, they diminish each other's power, perhaps even cancel each other out.

7. Physical and mental entrainments

Psychological archetypes are complex images. What is Father? What is Mother? Your immediate associations will indicate your current angle or viewpoint on those common archetypes in our human experience. That view will depend upon your own circumstances and history, generating its unique combination of constructive/destructive interferences.

A more destructive interference might manifest in your memory as a harsh image of Father or Mother. But the image might instead be constructively enhanced. Whether in Dad you see a wise Sage, crotchety Crank, or gruff old Softie—whether in Mom you see Madonna, a whore, or Madonna the Whore—whatever slant you hold will come from your own unique viewpoint in the universal hologram. Your own slant on an archetype creates your stance regarding it, and in a larger sense, it becomes part of your take on reality itself.

You can unify discordances in your psyche by integrating its dissonances and moving them toward more inner harmony with the flow of the Tao. You can adjust, even integrate some psyche dissonances just by changing your mind.

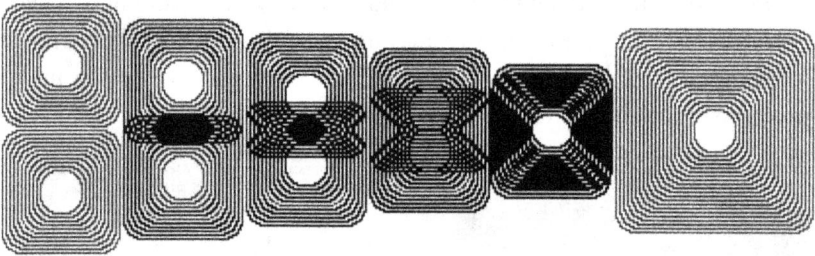

Two wave patterns merging into one

Above is a visual yantra for the integration of discordant energies. From left to right, two colliding patterns merge into one larger pattern. It suggests that colliding events need not disrupt your psyche; instead, they may establish a larger, more fulfilling unity. How? Energy travels in anything that can carry waves: air, water, metal, rock, even the electrical jelly of your brain. Those waves create crests and troughs in resonant relationships where one wave meeting with another wave can make no wave, a so-so wave, or a huge wave.

The ancient Taoists favored turning the chaotic waves of life to constructive purpose. To rephrase it into the laid-back jargon of Southern California surfers, the Taoists surfed the emergent curl of reality. They used the I Ching to ride on the crests and troughs of chaotic events in life's dynamic flow, following each swell, curve, and break toward the source of the Tao. They treated life as the ultimate wave, and living it truly well, the ultimate ride.

If you try to do this, it won't be easy sometimes. Keeping the right balance of straight-bore logic and cycling analog emotion to stay at one with the moving wave is a skill. Practice it daily, terrifying as that may feel in the calamitous curl of events that occasionally seem about to collapse over you in a wipe-out. Shoot on through that dark tunnel curving into the light. Find the right entrainment with your purpose to sync your energy into the flow of the Tao.

What comes to mind now is an entrainment that Christiaan Huygens discovered in his workshop in Holland back in 1665. He often gazed up at two clocks hanging from a wooden beam overhead to see that their pendulums were perfectly synchronized, swinging together in time, but in opposite directions, as if to balance out each other's motion. Huygens knew the pendulums hadn't been purposely set to such exact synchronization, and indeed, could not have been. So how did it happen, such total unison, he wondered?

Huygens said the two clocks revealed "an odd kind of sympathy." They must be transmitting their own vibrations to each other through the beam, he thought, bringing each other into synchrony through entrained vibration. "For a long time I was amazed at this unexpected result, but after a careful examination, finally found that the cause of this is due to the motion of the beam, even though this is hardly perceptible." Huygens had found mode locking.

Science has since discovered all manner of mode lockings, some of them quite uncanny. Women in a college dorm tend to cycle their menstrual periods together over time. In the night, mosquitos may start clapping their wings together on beat, or fireflies may start flashing together, or two lovers may start breathing in unison…and there are many other physical entrainments.

There are also mental entrainments. You can choose to become entrained in the universal flow of the Tao and let your life ease into more meaning. Changing your mind about something changes everything. The path of the Tao flows in the meandering watercourse way by natural gravitational pull. What seems like a mistake can turn out to be a bonus. Learning to move on the Tao's swift currents and slow eddies will reveal that life's apparently random events hold a deeper order. And not only an order, but even a unique personal meaning.

Discovering this natural order for yourself can resolve the nagging splits between objective and subjective, between discrete and holistic, between science and art, between matter and mind, between body and soul. You can find a transcendent third way via patterned co-chaos. It opens up a better future.

Some today suppose that we're probably binary images in a vast computer game played by an alien race. But that defines us in computer terms, forgetting that we mesh into an organic meaning fuller than computers can embrace.

Think organically. We live, and our universe lives. Patterned chaos is drawing the hard sciences of physics, chemistry, and astronomy into a broader stream of confluence with the softer sciences such as psychology and sociology, and even with the arts. These collaborations cross-fertilize each other to give answers that seemed unreachable a century before.

Science is beginning to discover a new connectedness, an interrelated, supportive web suggesting that the universe is more than just random chance. Science studied with a mystic's heart is now an emerging trend. There is scope for emotion here, for art, for passionate meaning—for all those squishy, vaguely "humanistic" qualities that science cannot account for, but can only count.

Interestingly, chaos theory has developed its passionate advocates. Scientists in a multitude of fields have turned their careers over to studying the multiple aspects of chaos theory, and for good reason. James Gleick said in *Chaos: Making a New Science*, "The first chaos theorists, the scientists who set the discipline in motion, shared certain sensibilities. They had an eye for pattern, especially pattern that appeared on different scales at the same time. They had a taste for randomness and complexity, for jagged edges and sudden leaps.... Believers in chaos—and they sometimes call themselves believers, or converts, or evangelists—speculated about determinism and free will, about evolution, about the nature of conscious intelligence."

Fractal nature touches the heart as well as the head. It gives an eye, an ear, a heart for pattern. A taste for complexity. In fact, some scientists experience a nearly religious appreciation upon finding this new window into reality, simply because it opens up everywhere and reveals its beauty so universally.

Converts to the beauty of patterned chaos may act as a new order of secular priests who honor a modern version of the Tao, where meaningful patterns trace out natural laws of quality as well as quantity. To recognize and honor both right quantity and good quality can bring the whole world into balance.

From the iterating clusters of resonance in society's behaviors come the various group values of legal, economic, social, sexual, ethical, and religious norms. Then linear logic takes over and tries to spell out and quantify what has already become a generally approved subtext in the culture, but which may eventually become so rigid in its dogma that authority hews its members to fit the rules. People make the rules, and then the rules begin to make them.

But reality always runs farther, faster, wider, deeper than the rules of society can quite manage to follow. Life outstrips our understanding continually in its rush to reveal truths beyond the codified norms. That dynamic comes from co-chaos operating in the living universe itself. Generated by the master code, its analinear patterns grew and still maintain this Double Bubble universe.

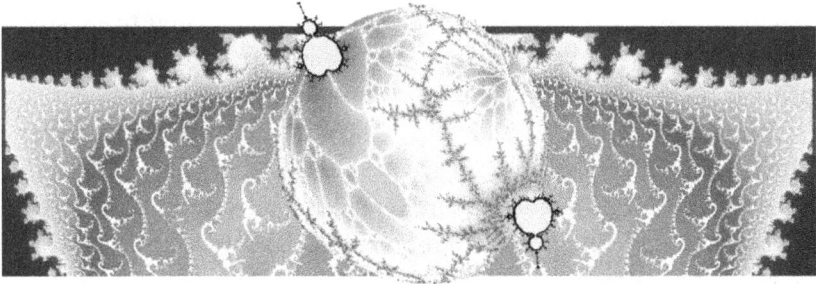

Chapter 6: Haunted by God

After that teen dream where God gently but decisively left me, I tried to shake it off as preposterous balderdash. Although, hmm, why did the presence of God at first carry me into such gracious ease and beauty, and then at the end, why did those burning words from God bring me to such aching loss?

1. Reducing God to smithereens

Why would a dream even matter? It was just a dream, I kept reminding myself. Obviously, it revealed my infantile longing for a divine super-papa to watch over me, no doubt because I'd soon be leaving home to fend for myself. Even though, at least consciously, I longed to get away from home. Yet somehow, that stupid dream showed up. So I apparently must be harboring some hidden anxiety about leaving my parents. How to quell that weakness?

And hey, why did that hunky young God look more like a lover than a parent? Oh, that's just because of adolescent sex drive. Okay…but why did God seem so tender and regretful as he left me? Oh, that's just to ratchet up the soppy tear factor. Don't go all sentimental, I told myself!

Wow! Why not use my new reductionist skills so recently honed on existentialism? My super-sharp teen ego stepped up to the logic-chopping block and hacked down that silly old God dream into tiny bits of childish wish fulfillment. Ha! Now the stupid dream was scornfully destroyed. I threw its confetti of ruined glory up into the air of my silent bedroom, and it came down labeled "Debris from a shredded adolescent projection." So there!

Looking back now, I realize that around age 12, I first began wrestling with the two issues of sexuality and spirituality. Simultaneously. They were the two poles of a creative drive that inevitably arrives with adolescence. It energizes the young, urging them to reach out toward that which creates…somehow. Maybe with the body, maybe with the spirit.

Older tribal societies initiated their young in ceremonies designed to promote a safe passage through the labyrinthine puzzle of adolescence into adulthood. But I grew up in a relatively jejune American culture that splays its teenagers apart on the horns of an adolescent dilemma: sex or sublimity? One path can degenerate into sex addiction; the other, into drugged devolution.

My newly honed intellect decreed, "The existentialists say we're no more than transitory flesh packs addicted to the illusion of meaning. Our lives are isolation cells. At odd moments, the meager conduits of my five senses release me from this existential prison of my body, where occasionally I may achieve fleeting bursts of communication with another prisoner celled nearby.

"Yet no one understands anyone, not really. There's no exit from this solitary prison called life. Except death. Life is hell, and then you die. For a prisoner locked in this indifferent universe, the only heroic choice is to adopt a rueful irony. Heroism is abandoning the illusion of God or any cruel joke of meaning."

Yep. Uh-huh. That's what I thought. Back then.

I was so sure of all this because Heidegger, Sartre, de Beauvoir, and Camus all said so...and they were approved by the reigning zeitgeist in post-WWII Europe that I so admired from afar—very far. Remember, I was sitting in Waco, Texas, reading Continental existentialism...not those zany American Beats who came along a few years later to play pranks on a bus trip.

The European existentialists were a glum, sardonic lot, full of woe-eyed angst. They had sallow complexions from living in coffee shops and dangling cigarettes from nicotine-stained fingers. They had spiky hair, black clothing, dead or fiery eyes. They drank espresso or absinthe, read despairing poetry. They looked blasé or enraged as they talked in smoky, dark lairs, alienated together in little cliques.

How did I know this? From *Movietone* newsreels narrated by Lowell Thomas and photo spreads in *Life* magazine. In far away Waco, Texas, I became alienated all alone. Who was I to question the psychic chill, wry loss, howling depression, heroic victimhood that shivered in such a mindset, hell-bent on isolation in an indifferent universe? How was I to know that by setting myself resolutely adrift in the world without trusting something greater than myself, I was booking my travel itinerary into adulthood with decades of anxiety, loss, and sorrow?

That summer I turned 17, I kicked God out as a fairy tale outgrown. But God had other plans. He walked into my dream that summer night and touched my left arm—the unconscious, unfocused side that links to my right brain (which would someday finally recall that long-forgotten dream, remember its promise, get the big picture at last)—and God said, "I am leaving now."

Not unkindly spoken, especially considering the next words: "But remember, I will come back." God's touch branded into my dreaming flesh a promise to return. It made my body remember, even if my mind forgot. Perhaps that touch coaxed me through the coming decades of sorrow, loss, regret.

2. Only Bob listens

I tried to tell my mother about that stupid dream where God left in the kitchen the next morning. She resisted hearing it. As I spoke, she kept her head turned resolutely away at the sink. Afterward, she would not look at me for hours. Said nothing about the dream. Her frown conveyed that I was sacrilegious, transgressive, disrespectful of conventions about God. God was a taboo topic if not confined to church, piety, or a standard curse of "Oh, for God's sake!" or "Goddammit!" Except cursing was forbidden in our house.

When my senior year of high school started a few weeks later, I began telling the dream to a new audience, my few friends there. But why? The dream was obviously silly, just a ridiculous dream…that I nevertheless could not seem to dismiss or forget. Not yet.

My adolescent friends didn't show much interest in it, though, which again made me wonder why I felt so compelled to keep telling the dream to people? Only my odd friend Bob listened when I talked about that strange teenage hunk of a God who broke my heart—at least in the dream—when he left.

Together we searched for a rationale as to why I would have such a silly dream. I trooped out all the existential arguments I'd read—each of us is trapped in a lonely cell of isolation, locked down to die in the remote chill of each other's gaze. Bob and I bent our heads together over alienation and became a clique of two discussing honor, authenticity, illusion in the midst of nothing.

I admitted to Bob that…um, maybe the dream was just a natural, inevitable part of being a teenager. Maybe I was afraid to leave home. Maybe the dream really meant I was just looking for Dad in all the wrong places.

Bob nodded. "I know about that." We agreed to scribble poetry for a while. My poem was about a young boy slogging along a sidewalk, looking down with a fixed, melancholic gaze, pushing his feet through the air like it was solid glass. Isolated.

Weeks later, Bob came over and asked to see again that poem about walking through air like it was solid glass. He kept thinking about it, he said, but by then I'd thrown it away. Just a meaningless poem in a meaningless world.

At that time, Bob was still just Bob, drifting through high school, far out on his own fringe, even taking several years of a Baylor drama workshop conducted for teens right across the street from my father's tuxedo shop. I could look out and watch Bob leave or enter Baylor's Drama building at odd hours.

As for me, I married impulsively at age 18. Why not? After all, I was psychically adrift, frustrated, poor, and it was Baptist Waco. Given Freud's two fascinations—sex and death—the former option seemed preferable to me.

As a wedding gift, Bob gave me the oddest pitcher I'd ever seen. White, curvy, but I could not serve anything from it because the snub-nosed spout was too truncated; it dribbled instead of pouring. Yet people could not stop staring at its odd silhouette on the gift table because it was so strangely beautiful.

Then Bob moved away and became Robert Wilson, stage director and playwright in New York City and abroad. We saw each other far less. He was so high-flying that he seldom lit near me. A renowned author-director in avant-garde theater, he made better use of Continental existentialism than I ever did.

For a while, I still kept trying to talk about my haunting God dream to anyone who would listen. But why did I even bother when it sounded so weird to people? I just knew the dream had hit me so hard in inexplicable way, but I couldn't figure out why. After all, it was just a silly dream. Of God, yet!

But I knew there was no such thing! How stupid can a dream be? Hah! Yet why would such a silly dream carry such a gut-punch? Why did it numb my mouth to talk about it, even my mind to think about it? Yet I kept talking about it, kept reliving that dream's impact in my mind. Why did it leave me so forlorn? I did not know…especially since I was so tough and wryly existential.

Notice, I could not admit to the dream's poignancy, nor why it kept affecting me, haunting me despite my bravado. I shrank from its uncanny clutch on my emotions. I did not want it to touch those feelings. When God walked away, I sought some quick put-downs for how and why it happened.

Now I can see my logic-chopping ego reduced that dream right down to smithereens, smashed it, trashed it…while overlooking its clear and simple message: God left me. At the time, it didn't even occur to me that I'd left God first. I'd replaced God with a bleak existentialism. Then God left me graciously.

I guess I couldn't face the significance of it, but something kept drawing my inner gaze continually back to that moment of goodbye—to the balm of God's gentle touch on my left arm in departure, laid square-bang against the agonizing burn of his words. That instant kept pulling me back like a magnet.

3. Driven to write

Even though I put up barriers, for months that dream kept pulling me back into an emotional riptide. Oh, phooey! Here I was acting like a dumb teenager, stunned after a break-up with some gorgeous hunk. Except this hunk was God, whom I'd never met in waking life. So how come his goodbye hurt so much? Why did it leave me feeling so raw? Because it just did, at some level below logic.

Oh, use logic, for heaven's sake! Why would that dream matter so much to me? Okay, it must have something to do with the innate mythic cast of the human psyche. Maybe it's about how we need to hear a story, tell a story….

Yes, of course! That's it! That's entertainment! Ta-dah! It's what puts the *force majeure* into movies, literature, theater. But why do we spend so much time and money on imagined ephemera? What exactly makes it so potent? Why do our emotions bend and sway to a story's words, thoughts, and images?

Hmm, if that dream just spun a mythic tale, then its absurd impact might lose force if I tried writing it down. Weaken its mystery. So I sought to tame that God-dream in study hall by inking it into the confines of blue-lined theme paper in my 3-ring binder notebook...and it became an addiction. For months, while my family watched TV in the evenings, in my bedroom I'd try to write out the dream, trying to rig a harness of words that would tame its power.

Have you ever tried to tell an amazing dream so completely that you encompass every nuance beyond words to tell? Trying shows you it's impossible. I could not subdue that dream or control it. I wrote it out so many times, in so many ways, that by January, I coaxed Mom to pay to let me take a night-school typing class downtown. Why? Because I had to nail down on paper that weirdly hypnotic dream. Tired of so much effort, I had to find more economy of motion in writing it out so often. So two nights a week, Dad drove me downtown to a little business school located above a loan company.

I had to take a night-school typing class because at Waco High, college-track kids could not get into typing, bookkeeping, geography, or Spanish classes. Such were deemed too blue-collar for us college-bound students; instead, we were routed into trig, Latin, British lit, physics classes. No joke, that's true.

Starting in January, I clacked away on a manual typewriter twice a week in night school. When the class ended after four months, I begged my bemused Mother to buy me a pawnshop typewriter and a ream of paper.

She did, bless her heart. But I still could not get that dream corralled into adequate language on paper. I even switched to a more expensive onionskin paper, so I could at least erase the wrong words and squeeze in new ones without retyping it all again and again in an endless, hopeless effort to tame that dream.

Surely you can see by now that I'm talking about obsession. That dream was driving me...somewhere. I did not realize it was teaching me how to write. I certainly did not question why. I only knew that I must get the gut-punch of that dream articulated somehow. If I couldn't do it by speaking it, then maybe I could manage to do it on paper. Typing on onionskin paper was much easier than just scribbling out the words again and again in inky longhand.

But no matter what, the clout of that nutty dream still stayed beyond my ability to capture it on paper. Or even convey aloud. Or even think about it much, I decided by the next summer. For maybe another year, that dream inexplicably caught at my imagination occasionally and squeezed. But it finally just dropped away, lost out of sight behind the burgeoning pile of new projects.

With this sudden new interest in writing, I launched into short stories, poems, speeches, plays, essays. The God dream was the jolt that galvanized me into a lifelong love of finding the right word...of words that ring true, or as true as I can make them. I still feel that pull as I finish this sentence and turn it at a new angle for a better view. Okay, good enough. Go on.

Writing is a solitary task. Right up my alley. Since lonely honor is the badge of the existentialist, then solitude while writing might be just the ticket for me. I viewed Hemingway as a model, except for that macho part. Then he killed himself in 1961. So macho. Well, I liked John Steinbeck, too.

4. Embracing existential angst

When my search for the right religion or ethics dead-ended in that stupid, meaningless dream about God leaving me, my new plan was to go secular. I would become the best person I could, for without God, honor is all we have to lift ourselves above the clay. Despite a strange, aching loneliness in the clamor of others, I mostly avoided the self-medication of alcohol, tobacco, drugs.

But soon, I did discover a new avenue for getting beyond myself. Sex. Actually, it was rather brave of me, too. I took up with a high school friend. He was smart, good-looking, quirky, with the added merit of a father who taught geology at Baylor and could wangle us into geology field trips around the area.

We'd go rocking. That didn't mean dancing. It meant looking for rock specimens of different kinds. So if I'd rejected a spiritual connection with God, at least I had a physical connection with someone who enjoyed nature, including sex, and who accepted it at a level below words.

That part really intrigued me. No words. We were novices at sex, but it was a strong connection, intense and affectionate. I was glad he was funny and interesting, and we spent a lot of time together. For a while. Then I veered off to seek a more alienated, unfulfillable romance—something with a lost God overtone—and he veered off to seek less smarts and more compliance.

When religion failed me, sex offered an emotional power and intensity that I'd craved to find in religions, ethics, or whatever else would give meaning to my life. Sex fired up my body, yet my soul drifted off toward inconsequential hell.

As I kicked out the divine possibility, something retreated into the unconscious where it loomed in vaguely monstrous shadow. An energizing force dropped away from my attention and lurched off to appear again in negative synchronicities beyond my conscious control. Evil is good warped out of true. Bedevilment is the power of God seen in a viciously distorting mirror.

Over time I discovered that I attracted some guys by pinging a longing in their souls as much as in their bodies. I connected with those inchoate desires and found an intensity of experience that I craved. I loved what those men revealed about themselves, about myself, about the startling juiciness of life.

But when a marriage ended despite my hopes and last-ditch efforts to hold it together, I found myself at age 29 divorced, with two small children to care for and no child support. I felt a failure, unable to hold onto a husband who'd left me to explore his latent homosexuality. I didn't know how to support my children with just a high school degree and not even minimum-wage job experience.

Feeling lost and valueless in every way, not knowing how to provide for my children, I decided that my parents would do a better job of raising them than me, and I tried to commit suicide. But what the hell! Even that failed. I woke up in a hospital room, delivered again right back into my skin and troubles.

I sighed, "I can't even do that right." Yet I also sensed that I would not go there again. Okay, if I could not live better for myself, maybe I could manage to do it for my children. And I tried. I got a Bachelor's degree, a Master's degree, a Ph.D., all the while working to keep our little family going.

I struggled with finances, career, health, and relationships. I achieved quite a bit in the everyday world, but it didn't fully satisfy. Not enough, not finally. Slowly I realized that despite my good intentions, willpower, intelligence, humor, resourcefulness, plus some racial, national, and cultural advantages... something was missing in the grand scheme of things.

So here I was, favored with a good education, good children, and good intentions! I lived in a land of peace and plenty. What was lacking? Career? Family? Friends? Money? Sex? Distractions? None of the above, yet it finally was all somehow not enough for me. Lovers or husbands did not fill it, nor did jobs, co-workers, friends...nor even my children and pets.

All of it was still not enough to sweeten a bitter taste lurking somewhere behind my life. I wanted something that was bigger, deeper, wider, wilder. Something true, not delusional. Something that was wrapped in the deep heart of mystery. I held an irrational longing for what was too marvelously beyond my scope even to imagine, much less name. Certainly not God. But if not God, what? A rueful "What?" became my big, nagging question.

All the while, I did not even realize I was being haunted by God, or rather, by God's absence. I did not understand that a loss of the divine was what ailed my valiant, struggling efforts to survive honorably in a world without God.

Trouble—meaningless, pointless trouble—hounded me all through those long years. Me, a basically good, well-meaning person. The problem was that nothing had my back, where the shadows lurk. I longed so deeply, unconsciously, for transcendent meaning past the dead ends of those courageous corridors of ratiocination that led to...nothing.

Meditation helped a little. Yoga, more. Friends, a lot. Eventually, after God finally came back, I discovered that the planets, stars, animals, and plants all maintain a sort of awareness that does not turn away from God. On this Earth, for instance, only humans turn away from God. Only we have enough free will to choose so consciously how we will live out our lives. Even animals, whether free to roam or not, mostly respond to the divine wonder in nature.

But now we humans make our own environments, mostly urban, and we mostly kick out the divine mystery to make a hell of it for ourselves and each other. We do it by turning away from love, away from sharing our best with each other, away from taking care of this world that we now have the power to destroy.

I eventually saw that love is what holds our universe together, and God could turn off the binder at any moment—but does not, out of love for all these creations, even though some willfully turn away in indifference or worse. The anguish of a parent at a wayward, ungrateful, self-destructive child is null compared to what God experiences when the created disown their Creator.

Yet we do not empathize with God, whose solitary quality of eternal existence alone, without parent or peer…this God trying to create worthy companions out of those who willfully refuse to grow wise enough to become loving, constant, and aware friends of God.

Yet we do grow, and it often involves learning to transcend a paradox. In this particular universe, we live by differentiating polarities and then transcending those polarities again and again into higher levels of resolution. God waits to greet us in the interstices of paradox. When two poles of a dilemma reconcile into a paradox resolved, for a moment, you can spot the fleeting face of God. For a moment, you get a brief slanting glance on that which contains all polarities and reconciles all opposites.

But God is not just the be-all and end-all. God turns a personal face toward each of us all the time. If you can return that sentiment, you will find divine mystery inside your work and arguments and sorrow and shocked laughter. As you realize that, you can open up to the wonder of the unfolding moment, and you can become a friend of God.

Or not. "Friend of God" is a term used by many cultures to signify God as something other than a lord or ruler or punishing parent. It means entering the shadowy unconscious and finding—shock!—that what had seemed like a fierce oppressor is the wise friend, lover, companion who knows you already. Every moment is when to begin. Every place is where to start.

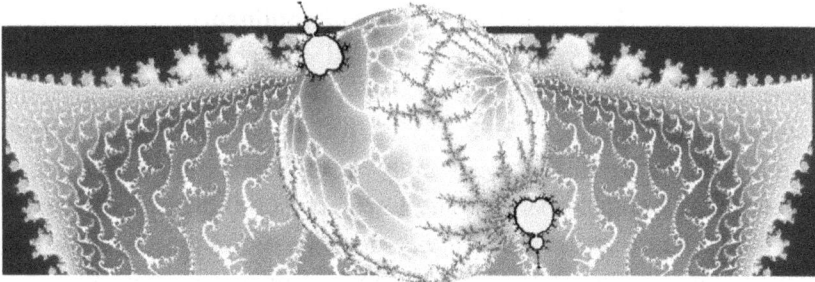

Chapter 7: Your Life is Nonlinear & Fractal

1. What is the importance of chaos theory?

Chaos theory is arguably the 20th century's most important scientific discovery. Perhaps you retort, "Oh no, Einstein's relativity theories embrace the paradoxical relationship of space and time." Or you might point out, "Heisenberg's Uncertainty Principle confronts the paradoxical relationship of matter and energy." Okay, I get that. But this TOE says that chaos theory encompasses all four of the universe's great primals—space, time, matter, and energy—allowing them all to work together as a polarized pair of pairs.

Consider the events in your own life. What do they share? They all exist in the spacing and timing of matter and energy. Ever-emergent events manifest their ongoing flow in ways that modern science can already often describe as deterministic chaos. Its hidden patterns appear at every level, from macro to mezzo to micro. They account for all sorts of previously mysterious fluctuations.

Here are some of the systems that exhibit regular chaos patterning at work: the cycles of epidemics, the rise and fall of stock market prices, the formation of hurricane systems, the beating of a heart, the rise and fall of white-blood-cell counts, the flux of caribou populations in the Arctic Circle, the eddies in a tidal pool, the smoke curling upward from an incense stick, the rise and fall of the Nile, sunspot cycles, the swirling gases of the Great Red Spot on Jupiter.

The airways of your lungs are living, branching, fractal shapes. A neuron mimics the fractal shape of a Mandelbrot heart. A fractal coastline shows the same grainy ruggedness at every level of resolution. Leaves have fractally determined shapes. So do snowflakes. Feathers. Blood vessels. Rivers. Fractals exist in all of nature, and they can be described by a mathematical science that charts the chaos patterns in life's nonlinear events on scales large and small.

All these fractal iterations of different basic patterns—faces, rivers, mountains, and skies—are what Hany Farid at Berkeley called life's *textures*. He is an expert in image analysis and digital forensics. He said he examines life's textures to detect digital forgeries and fake news. His techniques can spot hidden messages, track child pornography, test for scientific integrity.

2. What is the meaning of nonlinear?

Most of the events in life are fractal and nonlinear. But *nonlinear* is a deceptive, elusive word! Mathematician Stanislaw Ulam said calling something *nonlinear* is too vague a description. He said it's like calling most of the animals in a zoo non-elephants. That doesn't really explain or define much. Just as most zoo animals are non-elephants, so are most of the world's dynamics nonlinear.

You'd perhaps suppose that *nonlinear* means a system or process or equation is not linear. But, no, it only means that the system or process or equation is not *just* linear. It is linear-plus. Aspects of it are also analog, which affects the linear, turning it unpredictable. Now, a merely linear system stays predictable. If you push it twice as hard, you get twice as much response. Three times the push, three times the response. Its system is shifting proportionally.

And hey, many kinds of daily problems are linear. For instance, McDonald's can weigh one truckload of potatoes and estimate how many bags of french fries it will yield. Ditto for 10 truckloads. The propagation of sound waves is also linear. For example, if you shout down a well, the resulting echo is linear, because if you yell into it twice as loud, the echo is twice as loud.

But nonlinear systems are not predictable like that. Instead, their results shift disproportionately. Further, you cannot disassemble a nonlinear system into its parts and reassemble it back into the same thing again. Moreover, throwing an external factor into it will not affect the result additively, predictably. Instead, your result may get far better, stay so-so, or even become a Frankenstein monster of a result...like when a cheap amplifier, driven too hard, suddenly squawks into a shockingly raucous distortion of sound.

3. SKIP THIS SECTION IF YOU HATE MATH!

Equations map inputs to outputs. In a linear equation, the inputs and outputs stay proportional; in a nonlinear equation, they do not. In both equations below, **x** is the independent variable (mapped on a graph's horizontal axis), while **y** is the dependent variable (mapped on its vertical axis).

1. In this linear equation, $y = x$. *2.* In this nonlinear equation, $y = x^2$. What makes the second equation nonlinear? Just that ^{2} on its **x**.

Thus, a quick question spots the commonest kind of nonlinear equation: "Does its independent variable **x** have an exponent?" If *no*, the equation is linear. If *yes*, it is nonlinear because its output is *not* proportional to its input.

There's also a longer test to verify it: Linear $y = x$ will take a number value for **x** as its input and produce a proportional number value for **y** as its output. So if the input of **x** is **1**, then the output of **y** is also **1**. And if **x** is **2**, then **y** is also **2**. And if **x** is **3**, then **y** is also **3**. Thus the outputs stay proportional to the inputs. To check it, add all inputs: **1 + 2 + 3 = 6**. Ditto for all outputs: **1 + 2 + 3 = 6**. Result? The sum of inputs equals the sum of outputs: **6 = 6**. On an **x-y** graph, these inputs of **x** and outputs of **y** rise on a smooth diagonal line, meaning the equation is linear. (That's even how the term *linear* originated.)

Nonlinear **y = x²** will take a number value for **x²** as its input and produce a disproportional output for **y**. Okay, if the input of **x** is **1**, then **x²** is also **1**, since **1²** is actually still just **1**, making that output of **y** also **1**. But if **x** is **2**, then **x²** is **4**, making **y** also **4**. And if **x** is **3**, then **x²** is **9**, making **y** also **9**.

Now, check to see if the outputs stayed proportional to inputs. Add all inputs: **1 + 2 + 3 = 6**. Ditto for all outputs: **1 + 4 + 9 = 14**. No! Here the sum of inputs does *not* equal the sum of outputs: **6 ≠ 14**. The rising line on an **x-y** graph must curve. Why? Because the **x** does not have an implied exponent of **1**.

4. What is the main idea here?

The main idea above is that equations map inputs to outputs. Linear equations do it proportionally; nonlinear equations do it disproportionately. To quote Wikipedia, "Nonlinear systems may appear chaotic, unpredictable, or counter-intuitive." Talk about throwing a wild card into sedate, predictable numbers! Such tumult! That's why since early Greece, on down through 2,500 years, formal Western thought mostly concentrated on working out the linear problems that preoccupy us…for instance, 2 sheep + 2 sheep = 4 sheep.

People chose to tackle the linear problems for a very good reason. They were not only important but also solvable. But those far more numerous nonlinear problems in everyday events? Such puzzles were just too hard to solve, or even to perceive well enough to grasp mentally, so people stuck to the far fewer, strictly linear problems that offered readier solutions.

Before computers, the nonlinear realm of life was pretty much a mystery. The cycling, evolving shifts were—wow!—just impossible for our math. Mitchell Feigenbaum, a pioneer in chaos theory, said when he first started working beyond the fringe of linear respectability in the early 1970s, he was warned that even trying to comprehend a nonlinear system was too frightening.

But computers changed the ways we work with math. Now they can spit out the tedious iterations of nonlinear equations that would drive a million mathematicians crazy with boredom and overwork. We can at last enter the nonlinear realm in its huge, buzzing messiness and networking connectivity that makes any nonlinear system weirdly "sort of predictable." Why? Because its chaos patterns give a repeating general form, yet the evolving variants of that form have unique details. Why? Because along with having at least one constant, it also has at least one variable producing the continually shifting details.

5. What are chaos patterns?

Scientists investigate chaos patterns from two directions: moving from the simple to the complex, and from the complex to the simple. For our purposes...

1. **Chaos Theory** says simple things can generate complex outcomes that cannot be predicted by looking at their parts.

2. **Complexity Theory** says complex systems can generate simple outcomes that are not fully predictable.

Both approaches study the dynamics of chaos patterns in action, so I tend to say *chaos patterns*. Or sometimes *patterned chaos*. Or maybe *chaos patterning*. But there's a built-in mental snag here: how can the word *chaos* even sit comfortably next to the word *pattern*? The notion is a paradox!

The words *chaos* and *pattern* team up to mean that a nonlinear dynamic is deterministic and predictable in its general form, yet it also iterates with varying contents of specific details. Like weather does. Like your daily life does. Like the behaviors of your mate, child, and dog do.

In weather, chaos patterning sets the general form of a tornado, but each real-tornado iteration has its own unique, specific details. Ditto for snowstorms or thunderstorms. That's why scientists can make a general prediction for a hurricane, but they cannot reliably predict its exact strength, path, or duration.

In fact, the global weather is such an extremely complex system that it holds a multitude of chaos patterns. Weather is so inherently unpredictable that its vagaries were dismissed for millennia as "Acts of God."

Edward Lorenz, a math professor at MIT, helped to change that. He began to make weather more explainable as chaos patterns in 1961 when he used a computer to track 12 different variables affecting the atmosphere. Lorenz was shocked to discover that it was nearly impossible to gather enough data to make any truly accurate weather predictions, due to an extremely sensitive dependence on the initial conditions of measurement.

Lorenz found that a hidden mathematical structure shapes Earth's atmosphere. It is a vast system of underlying chaos patterns. "Even with vast amounts of information, reality would inevitably drift away from the forecast," explained Adam Kucharski in *Forecasting the Chaos of Tornadoes*. "For something to meet the mathematical definition of 'chaotic,' it does not just need to be sensitive to initial conditions. It also needs to have an underlying structure."

Lorenz's work on the chaos structure of weather swung open the door to a new science: chaos theory. Back in 1887, Henri Poincaré in France had already cracked the door open a bit to see chaotic deterministic systems. Still, with no computer available, he could not handle the exhaustive computation of huge numbers of shifting details, so Poincaré could not take his idea very far.

The hidden realm of chaos patterning began to appear only in the 1960s when computers started to chart those nonlinear dynamics with great sensitivity to initial conditions. Physicist William McHarris said in 2009, "Chaos theory has hit main-stream science only during the last three decades, and it promises to change not only the way we do science, but also the way we think about nature in general. For nonlinear behavior is the norm, not the exception—this means that most of our neat linear models break down and are not really applicable." Patterned chaos has its own special signature:

> ❧ 1. Nonlinearity
> ❧ 2. Order beneath apparent disorder
> ❧ 3. Cause & effect are not proportional
> ❧ 4. Cycling that repeats in slight variations
> ❧ 5. Scaling sizes like nesting boxes
> ❧ 6. Universal applicability

Characteristics of patterned chaos

This new science verges on visual art in its gem-like fractal graphics and on poetry in its evocative landscape of verbal images: *strange attractor, butterfly effect, antennae, Julia set, Koch snowflake*. Patterned chaos lets us see an order hidden in nature's apparent random disorder. We rise to a higher level that discovers simplicity can generate a complex, unpredictable flow. Yet paradoxically, complex details can also camouflage an overall simplicity in nature, where the wonder of "sort of predictable" nonlinear patterns continually emerge.

I Ching shorthand uses a special kind of nonlinear math that I call analinear. It can chart big patterns in nature's flow. Its simple, irreducible core employs a polarized pair of pairs. Take, for instance, the master code that creates the four great primals of our universe: space and time, matter and energy. It generates and projects the hologram of our Double Bubble universe.

The ancient Chinese saw this special patterning in nature, including human nature, and they termed it the nesting boxes of the Tao. Chaos patterns in nature sculpt events in matter and energy over space and time. The purpose of Taoist study was to discern how to live wisely in this weirdly "sort of predictable" flow of events, and they used I Ching math figures to chart it.

Confucius was said to declare in the *Confucian Analects*, "If some years were added to my life, I would give fifty to the study of the I Ching, and then I might come to be without great faults."

According to Szuma Chien in *Records of the Historian*, Confucius read the I Ching so often that the thongs holding his bamboo-slat text together broke three times. However, some scholars insist it's unlikely Confucius ever saw the I Ching, and others of humbler stature ascribed their I Ching insights to him.

6. What is your daily nonlinear experience?

How do you parse out logically the rhythm of a 4-year-old whirligiggling on the lawn? Or a wave curling against the sunset? A bird song at daybreak? You stand there and watch and listen, and you walk away with a memory of what will never again repeat—not exactly, not in precisely that way in that space and time. Even with a camera, you cannot hold onto the real event. You can get a stale facsimile. A memento. Sure, you can pick up a shell or feather, or you can take a photo or video to commemorate the moment. But you won't carry off its whirl or wave or bird—only the feather, photo, recollection.

When we stand on the lawn and watch a child play, or we pause on the porch to listen to a bird's song, we experience this exact moment's unique variant, not an end product. We are transfixed by an iteration rather than rushing to a goal. We go with the flow. We participate in change instead of controlling it. We are surfers riding the endless wave rather than beachcombers carting off bits of dead matter to memorialize dead events. (But I still like beachcombing.)

Sensation is always conveying life's details to our awareness via shifting analogs of quality, rather than crouching down to measure it out using the linear units of quantification. We do not think, "Ah! 780 nanometer wavelengths of light. Logic tells me this is red!" Instead, we experience it as the shock of blood flowing, a toreador's cape, a stoplight, a lipstick kiss on the cheek.

And due to sensory saturation, the mere quantity of something becomes meaningless after a while. We get deadened to it—whether it is the national debt, number of bedmates, parcels of real estate, or bottles of beer.

In co-chaos, a chaos dynamic (already pretty powerful and dependable in itself) is teamed up with another chaos dynamic. They bond into a dynamic duo far more powerful than either one acting alone. Co-chaos is much more stable and failsafe than mere chaos. It maintains greater security for the basic form, while it also iterates a flow of evolving contents that are self-similar yet ever-unique in their details.

Co-chaos is stable yet changing, safe yet adventurous, securely itself yet constantly evolving. Much as you are. Much as the universe is. Your DNA is based on a co-chaos paradigm that iterated a specific variant of the species: you. You live your unique life, yet you're also inherently engaged in a feedback loop that exists to maintain and evolve your species and its relationship to your world.

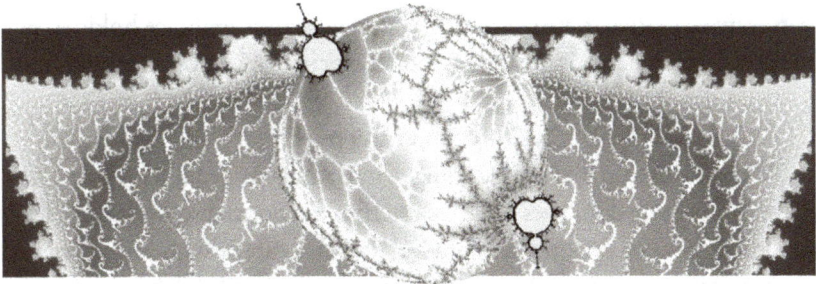

Chapter 8: Divine Chaos, Orderly Litter

1. When God came back

In a 1958 dream, back when I was 17, God left me. Starting right then, I began to live a long season in handmade hell. I had more pointless troubles than I could keep count of, so I won't try to enumerate them here. Despite my guts and grit, it was the daily grind that turns you into ground meat.

By my late 30s, though, my identity became more than just my endless tasks, thoughts, emotions, or reputation. I began to embrace a new awareness that I could not exactly define or explain. For me, finally now a university teacher, it was a new attitude—but how I did it or why life opened up remained unclear to me in ordinary logic.

Then in 1981, affable, rangy-framed John Walter entered my personal life and transformed it with his good humor and easygoing kindness. By the time we eventually married in 1984, I was living in a routine of ongoing happiness. Every day, no matter how bad, was also good.

In 1985, God came back in another dream. I wasn't expecting it. I'd long ago dismissed and forgotten that teenage dream where God left me. But after 27 years, God suddenly showed up in a dream one night. It was a dream of total union that took me back to the beginning of this universe. I flowed back through its reversing hierarchies of creation to the universal startup.

Then I went on beyond that to what was holding all the spawn of universes together, something bigger that encompassed all of them and still more that I could not fathom. It carried me—me, an unchurched, humanistic doubter!—into a place where all things merged in a huge unity beyond nothing.

Then I became one with that unity. It was ecstasy that pales sexual orgasm to a whisper. I felt the passion that was holding together all the universes, and so much more that I could not discern or understand. It was all held in an embracing love so profound that I could only gasp in awe.

In that encounter beyond belief, I was one with the All, and I knew its love beyond knowing. It encompassed universes and other things that were strange beyond my limits or definitions, for the Grand Organizing Design that we personify as God is so much more than universes. Its embrace holds all in a purposeful design that is more than we can ever know with our human minds.

2. Children of the psyche

When I woke up the next morning in my familiar bedroom, I was stunned. Alive in my own skin, eyes open, registering the normal sensory data. I did not know how to respond to that God dream…except with love. I wanted to share with others what I *saw/was* in that dream, but I didn't know how to declare such shocking simplicity. I could not figure out how to bring its treasure back into the commons of everyday life, ready to share with anybody who wanted it. Yet I also could not walk away from trying to do so. Within a week, I realized—delicious agony—that this was my task: learning how to share it.

Three months after God came back on March 4, 1985, John retired from teaching at the University of Texas. I told him I wanted to go to Zurich to study what dreams mean symbolically. The dollar was up; the barriers were down; he was ready to roam. John said, "All right, I've done my career. Let's do yours."

So I resigned in June from my job teaching at the University of Texas, and we both moved to Switzerland by September. By day, I took classes at the Jung Institute in Zurich's suburb of Kusnacht. Several evenings a week, I also studied in the library of ETH, the Swiss equivalent of MIT. At home, I read science and I Ching books and ate John's often-gourmet cooking.

This routine was stimulating. The Jung Institute helped me explore the mystery of dreams, especially my own. The ETH helped me study the genetic code as I tried to decide if its math correlated with I Ching math. I found myself living a better mystery story than I could ever write, for it was not based on why a person must die to set the plot going, but rather, on why we must live.

Then on October 30, 1985, I awoke from a silly little dream…

I am giving birth. It doesn't hurt. They pop out so easily. The first one is a huge boy, over 9 pounds. His name is Reuben. The second is Rachel. The third is Rebecca. Then comes a rush of three smaller children with no names…like runts of the litter. Six children in all. I smile. It's sorta like the way a cat or dog gives birth. So quick and easy that it's even fun. I'm laughing about it.

I woke up smiling at my silly little dream, amazed at the ease of all those dream births. The first three babies who came tumbling out already had names, and then out popped a nameless trio in succession. In that quick, absurd dream, I was cheerful and competent, easily birthing so many children.

Yet I'd actually given birth twice, so I knew it wasn't that easy.

Four days later, I walked over for my weekly analytic session with Elisabeth Ruf. All Jung Institute students were required to undergo 90 hours of analysis.

I chose Elisabeth Ruf because she had known Jung back in the day and was near the Jung Institute. I'd already seen her three times in weekly sessions.

I off-handedly reported to her that silly little dream of giving birth to all those babies. Frau Ruf looked at me shrewdly, nodded, and said, "Let's count back now. How long is it since your God dream?"

"It was this spring," I said. "On March 4, 1985."

She counted on her fingers. "That's about 240 days. It's the normal gestation range. At the low end, of course. But normal for multiple births."

I caught her drift finally. "No! It can't be!" I exclaimed. "That's absurd."

"What? You think you're not human? It is a human gestation period."

"What do you mean?"

"Intercourse in a dream usually brings forth something. Why else do you think God would make love to you?"

I floundered. I thought back. Good heavens! No, it couldn't be! Then I said slowly, "Uh…hmm. You know, right after that God dream back in March… that's when my menstrual period stopped, if I look back on it now. But I just supposed that maybe I was starting early menopause."

Then I exclaimed, "Oh! And my period started up again this week! On the morning of that dream of giving birth!" I stared at Frau Ruf, feeling naive as an ancient Greek girl caught unaware by a god in Mount Parnassus.

"These are not children of your body, Katya dear, but of your psyche. Your mind."

I said slowly, "Children of the mind?" I paused. "Okay. I guess the babies are born now. But I can't tell you much about them. Who are they?"

Ruf answered, "You don't know yet? Have you looked up the meaning of those three names? They are all Hebrew names. Do you know why?"

"No."

"Then go to the Jung Institute library and find out. Look in the name dictionaries for the meaning of their names. They do mean something, you know. Your babes have those names for a reason. Find out who they will become."

She paused and examined me with measuring eyes. "And do not let that huge baby boy, Reuben, take all of your energy. He is heavy. Nine pounds? He will wear you out if you are not careful. Save something for the other children."

3. What do their names mean?

Dazed, I walked away from Elisabeth Ruf's place over to the Jung Institute a few blocks away. In the library, I looked up the names of those children born of my psyche. All three names were part of the Judeo-Christian heritage and the collective Western culture that I'd been born into, grown up in.

Tears came into my eyes as I read what *Reuben* means: "Reward for love and prayer." I knew what it pinged for me. It recalled those adolescent years of searching for a God not locked up in stale words and automatic rituals, searching for something that would open our hearts and best thinking, where science and spirit can support each other, not deride and maim each other.

In adolescence, I'd found most of the Waco Baptists I'd met to be full of cant and can't. To me, it seemed their church gave God lip service, but it did not guide people to set their roots in truth. It gave them a social club.

And then, by the time I was teaching at the University of Texas, I was working around so many intelligent, well-meaning, clever people who treated religion like it was a cultural cul-de-sac, whether they were in or out of it. They acted like religion was something to be tolerated…or ignored…or removed… as if it was an appendix, vestigial and scheduled to disappear in social evolution.

Yet those same people wished for a viable code of ethics that faith no longer had enough strength to call forth in the collective. Perhaps Reuben was my reward for wanting more for me, for them, for us all…a merger between science and spirit, between cognition and the mystery of God.

4. Settling into the task

By March of 1986, I was making notes on correlations between DNA and the I Ching, using books in the ETH Library of downtown Zurich. I was still trying to figure out if the genetic code and I Ching shared the same math paradigm. Meanwhile, at the Jung Institute, I began to suspect that the human psyche also holds a lesser variant of that same code. Jungian typology describes it, and the Myers-Briggs Type Indicator standardizes it in a test instrument.

At the Jung Institute, I was also starting to learn how to read the messages sent to me nightly in dreams. My unconscious mind plucked them out of 3D time's greater perspective as my body lay asleep in 3D space. They helped me foresee and prepare for the next day, week, year. They set my feet back on the path of the Tao whenever my ego occasionally, inevitably, strayed off.

Hindsight showed me that a plan was deeply woven into my life's seemingly random details Looking back, I realized it was my insistence on linear logic as the high road to reality that had banished me from God. By dismissing a God I could not fathom, I'd limited the scope of my world and myself.

I realized that when God walked off from my teenage ego in a dream back in 1958, that hypnotic fascination of my internal gaze on God's goodbye had impelled me to learn to write. When God left me in 1958, promising to return, he'd parked my soul on a ticking meter of anguish. He walked away, and I did not know what to do with that moment except record it again and again, always unable to describe its power and impact, spending a long apprenticeship in chilly hell. Then I became fascinated with the writing process itself. But I must have been a slow learner because it took 27 years for God to come back.

So when God returned in a dream and showed me the Double Bubble universe, maybe he wanted me to write about it. Back when God left, I'd thought describing its inexplicable, poignant power was hard! But using an old Royal machine to type on thin, slick onion-skin paper and making frequent, smudgy erasures was a stroll in the park compared to this new task of verbalizing how a master code and transcendent love generated space and time.

How even to begin? Space and time began to haunt my reveries. Just beyond my ordinary vision, I kept seeing four bricks etched with yin and yang symbols. The four bigrams. They kept dancing in an eternal ballet that created space, time, matter, energy. Everything! Somehow?

By now, I was at last realizing how much space and time mean to me, how much we owe to them. Just consider. Space and time form the matrix holding whatever you do. Your life is a jewel—however rough or fine-faceted it may be—nestled in its own unique setting of space and time. Mine, too. So in casual conversation, why don't we talk more about this space and time setting?

We just don't. Do fish ponder the ocean they swim in? Do roots fathom the soil where they grow? We all just accept our own basal medium. This spacetime holds us so fully that we seldom bother to realize it, much less wonder about it, far less appreciate it, hardly at all talk about such a wonder.

Or if we do, what do we say? Vague approximations. Unless we're scientists who spout numbers. Or poets who weave metaphors. How to put the two approaches together? I realized that how we conceive of space and time shapes our comprehension of ourselves, our universe, and our God. How we conceptualize its invisible dimensions dictates how we navigate them. If we can change the way we navigate space and time, we can change our destiny.

Trying to understand what I saw in that dream and how to convey it, I studied at the Jung Institute by day and the ETH library by night. Meanwhile, I also explored yoga, *tai chi,* and many synchronicity systems.

On weekends I lounged around the house eating John's cooking and reading books. I read a lot of popular science, and in late 1987, *Chaos: Making a New Science* by James Gleick turned on a light bulb for me. Oh, joy! I'd already proved to myself that the genetic code and I Ching figures shared a common math paradigm…which I now realized might be fractal! But how to explore that idea and really understand it? Harder still, verify it?

5. Reuben gets mouthy; 3 into 2 won't go

By 1988, the third year in Switzerland, I was weary. I told John, "I'm so tired of carrying this load of I don't know what, going I don't know where. It feels like such a massive burden. I'm so tired." That night I dreamed...

I am carrying a baby boy on my left hip. He is big and heavy, huge. He has light skin, dark eyes, dark curly hair. He sits on my hip, looking at me calmly, expecting me to take care of him, as is his right, the way babies do. He gazes into my eyes and says, "I know you're tired. But you'll have to carry me a while longer. You know I can't walk yet."

When I awoke, I knew it was Reuben I'd been toting in the dream. My reward for love and prayer had just informed me that he would take his own sweet time growing up, developing his own two legs to stand on. And the manuscript just kept on growing. I recalled Elisabeth Ruf's caution that Reuben would be very heavy and demanding of my time and energy. I began to call the manuscript I was writing *Tao of Chaos* because that's how I felt about it.

The manuscript kept on growing. Maybe I should divide it into volumes? Then on November 22, 1988, I dreamed...

My three babies are in tiny white tubs that look like restaurant tubs of butter or jam. A woman takes one baby into another room to discuss news events with it as it rests on the padded arm of her rocking chair. I can hear their words clearly.

The woman is reading news to the baby. It answers, repeating her words and elaborating on them. She calls out to me, "Watch this. It's such an amazing baby. It can already talk about whatever I mention to it."

I'm in the other room, but somehow I know when the woman accidentally flips my tub baby off the arm of her chair onto the floor. At first it keeps on talking, so I think it's still okay. Then it begins to hemorrhage inside the tiny container while it is still talking. Then I hear it begin screaming. The container swells with blood. I come running into the room. Scared for my baby.

The woman tells me it is dead. Oh no! I mourn it. Then I look at my watch and say, "Ho-hum, time to get on with doing things—it is 5:44, almost a quarter to 6."

When I awoke, I thought, "Okay, in that earlier dream, I had six babies—Reuben, Rachel, and Rebecca...plus those three unnamed babies, who now may be showing up again as the three tub babies."

Did those unnamed tub babies refer to the overwhelming stacks of manuscript, notes, files I'd gathered? Just the day before, I'd thought, "Maybe I'll split the swelling manuscript into three volumes. No, I'll edit it down to two because I'll never understand how to describe the origin of space and time well enough. It's hard enough to do the genetic code-I Ching math correlations."

So one tub baby fell to the floor and hemorrhaged...maybe it died from swelling with so much data? Yet why was I so *ho-hum* about that injury?

That dream made me reconsider killing off one of the six books. Then I recalled the time in the dream: 5:44, nearly a quarter to 6. Maybe the tub baby swells with so much material that it will rupture into six books?

6. Reuben is hungry; Rachel gentles; Rebecca gives me joy

Oh, heavy Reuben! He just kept on growing, ballooning into computer files of text and graphics on DNA, I Ching, chaos theory, cosmology, quantum theory, gravitation, etc. I've felt Reuben's weight in my work, always consuming my energy and forever demanding more: "Feed me the I Ching, feed me the genetic code, feed me chaos theory, feed me cosmology, feed me string theory, feed me gravitation and spacetime latticing! Data, data, data! More, more, more!"

Reuben's influence probably shows up in the odd-numbered, more scientific chapters of this series. Nevertheless, his name means *reward for love and prayer*. Maybe Reuben's job is to open the hearts and minds of intelligent, well-meaning people to see how science and spirit can support each other instead of fighting against each other.

And Rachel? Her name means *gentleness*. Perhaps her influence is in the even-numbered, smaller chapters that are more philosophical and personal.

Rebecca, she of *enchanting beauty*, where does she show up? Rebecca gets her due when I explore symbolic imagery in various ways, for instance, in the I Ching text. And to take a break from tending Reuben's heavy data demands, once I spent a month writing *Dream Mail*. That book showed how to work with dreams. My relief from toting Reuben became playing with Rebecca. Such a dream girl she was, too. She just wrote herself, exploring my wonder at discovering dreams as personal stories with a point, promising, and emphatic... not just clanking symbols that clash by night to signify nothing at all.

Symbolic imagery does enchant us—to our amusement, sorrow, joy, ruin, and revelation. To understand more about the symbolic webbing stored in the unconscious, I've examined tarot, astrology, and other synchronicity systems. I even developed a card set of imagery that I call Kairos cards—part tarot, part Rorschach, part Thematic Apperception Test. I use them to help people tap into their unconscious wisdom via their projections on each card's image.

Those cards are especially useful for people who do not remember their dreams much. And when my dad had a stroke and could no longer talk, I used them with him. A few months into his new silence, Dad's eyes widened in relief as he pointed at cards as I laid them out, describing what the images suggested about his unspoken thoughts. His eyes teared up in gratitude, his good hand clasped mine in fierce thanks as the Kairos cards drew out his hidden feelings and thoughts locked inside that enforced silence.

Reuben, Rachel, Rebecca. Perhaps they are qualities, not books? And that tub baby who swelled and ruptured? Maybe among all the files, pages, and notes, it was referring to a 50-page paper I wrote as part of my final exam on psychopathology at the Jung Institute. In it, I described a gentler way to treat the beleaguered psychopathic psyche, not viewing it as fighting an insane war against society, but rather, as a treatable, correctable imbalance of the psyche's four functions of sensing, intuiting, thinking, and feeling…and its four attitudes of extraversion, introversion, perception, and judgment. That paper describes how the psyche can rebalance its four functions and four attitudes to discern reality more clearly and make better decisions.

But that paper got tucked away, and it died on the vine…or at least seemingly, for decades. But when I finally extended this series into Volumes 4, 5, and 6, parts of that paper got tucked into some chapters that describe how the psyche uses the four functions and four attitudes to discern the nature of reality more clearly…and how to bring them into better balance.

It's over 30 years now since *Chaosforschung* was published in the original German, which then came out in English as *Tao of Chaos*. Eventually, I split that original book and augmented it into the six-volume series, where you're now reading Volume 2, *Co-Chaos Patterns*. Six books. A sextet of dream babies. My life serves a plan I did not even know existed. But my dreams did.

During those 5 years at the Jung Institute in Switzerland, I worked on parallels between the genetic code and I Ching, trying to get that correlation done properly and published. It happened! My first book, *Tao of Chaos*, won a few small awards, maybe because all those charts of DNA, RNA, and I Ching math carried Reuben's weight. People even remarked on it. For instance, in 1994, when I first sat down in the consulting room of an Austin internist, he said, "I see from your file that you're an author. What did you write?"

I said, "I wrote a book called *Tao of Chaos*."

"Hey, I read that!" For several minutes, he questioned me about it. Meanwhile, he kept scrutinizing me, looking oddly puzzled. He finally said, "*You* wrote that weighty tome? But you seem so…soft." I think he meant ordinary.

Dr. Edwards apparently could not fathom how that "weighty tome" came out of soft, ordinary me. Well, isn't that the mystery of children? This book you are reading now carries heavy, demanding Reuben in its odd-numbered chapters. Rachel, she of gentleness, is in the even-numbered chapters. Rebecca smiles in the enchanting fractals strewn by the master code throughout the series. Those three children birthed a merger of science, spirit, and beauty.

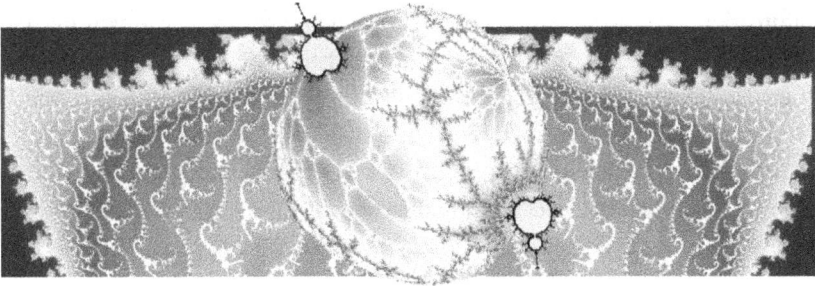

Chapter 9: Bifurcation

How do you create co-chaos? First, you make chaos. Patterned chaos, that is. Best way? Start with *bifurcation*. Splitting in two. Forking. "You take the high road, and I'll take the low road." "She lived on the morning side of the mountain, and he lived on the twilight side of the hill."

In ancient China, *yang* signified the sunlit part of a mountain that grabbed your attention, while *yin* was the mountain's shadowy area that got less notice. They symbolically contrasted assertive energy and receptive energy. Light and dark are also contrasted in the Italian word *chiaroscuro,* which literally means "light/dark." Here in a single word, you find showcased the polarized state of both opposites at once…bright, assertive yang and dark, receptive yin.

In bifurcation, a single behavior splits into two related yet different behaviors. Like this: suppose you're looking out the window, and you see a flag draped in lank folds on a flag pole. The flag is not moving.

But now the wind comes up, and you watch the flag flap. It collapses with a buckling motion. It flaps again. Buckles again. For at least this moment, the flag's behavior is alternating between flaps and buckles. It is bifurcated. This event is important. Leon Glass and Michael Mackey point out in *From Clocks to Chaos: the Rhythms of Life* that bifurcation is a universal trait of patterned chaos. Whenever bifurcation occurs, patterned chaos is likely to follow. That patterning is predictable in its general form but not in its exact details.

1. Bifurcation puts forks in the road

Okay, you're still watching the flag out the window. It exhibits bifurcation by forking from a steady state of lank inaction into two alternating behaviors that dance between either *flaps* or *buckles*…at least for this moment.

Of course, if the wind increases, the flag might start flapping briskly without collapsing into a buckle…or it might end its motion and drop back into the original static condition of lying lankly draped against the flag pole.

To chart the two behaviors happening right now, you draw a little bifurcation tree (b-tree) to represent the *flaps* and *buckles*. First, you make a dot. Then you draw it upward as a line representing the original steady state of the pole's lank, motionless flag. You count this single line or trunk as *Period 1*.

Next, you fork the trunk upward into two branches, shown here as *Period 2*, with the two behaviors of *flaps* and *buckles*. Both branches sit on a horizontal plane. This horizontal level holding the two branches in *Period 2* can be read across the b-tree as a *horizontal period 2 window*…or for short, just *hp2*.

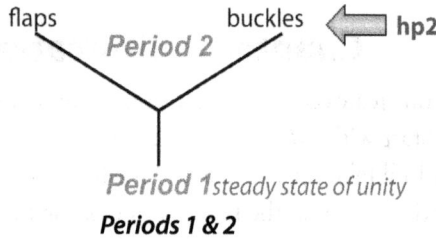

Periods 1 & 2

However, instead of graphing the bifurcation of flag behaviors upward as tree branches, you could also chart it downward into wells of attraction. The image below portrays the flag's action in this new way. *Period 1* represents the original steady state of that lank, motionless flag on the pole. Its lank droop exists in one static well, so there is just one attractor point. But on the right, there are two attractor points; *flap* and *buckle* sit in separate wells.

1 well of possibility has 1 attractor point
Period 1

2 wells of possibility have 2 attractor points
Period 2

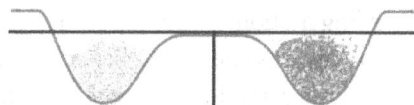

| **trunk** | **flap** | **buckle** |
| steady state of unity | attractor | attractor |

Wells of bifurcation

Bifurcation can graph the cycles of organisms. For instance, if a bird or bee population is stable, it exists in one static well as *Period 1*, where the number of births in the flock or hive consistently just about equals the number of deaths. Its population rate sits calmly in a single well of attraction. But the population may bifurcate into two wells or basins of attraction as *Period 2*. For instance, maybe the young die off from an illness, but the adults don't.

Biologists chart population shifts in all sorts of living things: weeds, trees, bobcats, deer, elephants, snakes, rats, fish, and all manner of insects.

Sometimes a population—of animals, insects, viruses, or even baking yeast—can develop a period-doubling pattern that keeps on forking up the b-tree.

For instance, in wheat bread dough, the traditional yeast used to make it rise will double every 20 minutes at 33° centigrade. As the yeast population doubles repeatedly, the dough keeps on rising...but only so long as the yeast continues to have food. When the yeast eats it all up, the bread stops rising.

That's why bread recipes will contain the carbs of wheat, honey, or sugar; the yeast needs to eat it while rising in a warm environment. By contrast, a sourdough starter that's kept in a refrigerator can stay alive for years by suppressing its appetite with a cool temperature and just kneading in a bit more flour occasionally to feed the yeast. I've done that myself.

An animal population can also keep on multiplying, given enough food and health. Witness all the begets in the Bible. Witness the human population statistics over the last two millennia. But it could instead develop the rhythm of a 4-beat or 2-beat cycle, or even drop back into a monotonous, steady state of just one static point in a single well, with the number of births and deaths in the herd, flock, or city staying just about equal. That's zero population growth.

Okay, look at what is happening now with our flag. As the wind grows stronger, the flag starts an intricate little dance. The flapping on the left branch elaborates into either a sharp *crack*, or alternately, a hollow *boom*. Meanwhile, the buckling on the right also begins to fork—into a *twist* around the pole, or alternately, a *shudder*.

The flag has now developed four different but related behaviors!

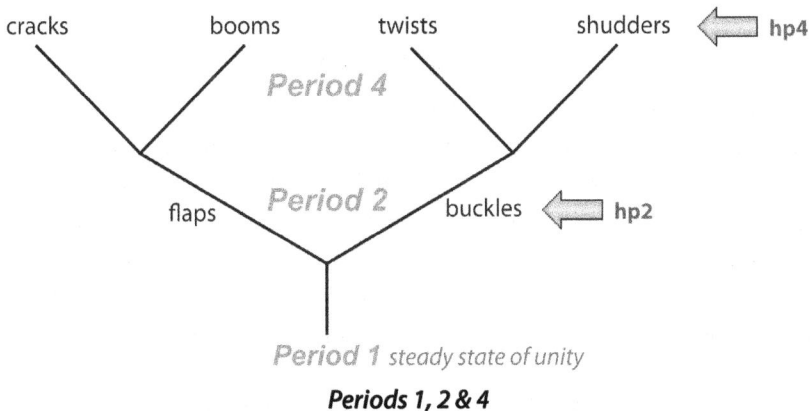

cracks   booms   twists   shudders   hp4

Period 4

flaps   *Period 2*   buckles   hp2

Period 1 steady state of unity

Periods 1, 2 & 4

So now the graph forks upward to show the four distinct behaviors in *Period 4*. The left branch of *flaps* has bifurcated into two higher branches of *cracks* and *booms*. The right branch of *buckles* has bifurcated into *twists* and *shudders*. All four behaviors interplay in a rhythm of *cracks, booms, twists,* and *shudders.* This higher level of forking is read horizontally across the b-tree as a *horizontal period 4 window (hp4),* since it holds four branches.

The depiction of this flag dance is a simplified event. Only a high-speed camera could register all the distinct, separate positions that occur when the wind moves a flag. Your eyes don't have stop-action vision, and it would all happen so fast that you could not easily discern all its distinct motions.

Thus, I'm just giving you a simplified bifurcation model here. It's easier to study movement in multi-colored liquids. They shift more slowly and visibly than air. Easier still to track slow-moving solids that may alter at a glacial rate.

For this dancing flag, let's double the behaviors once more by branching upward into a third level of activity. At this new, higher level of the b-tree, a horizontal window across the tree holds eight different periods of related behaviors: *snaps, beats, bams, pops, hums, crinkles, shivers,* and *shakes.*

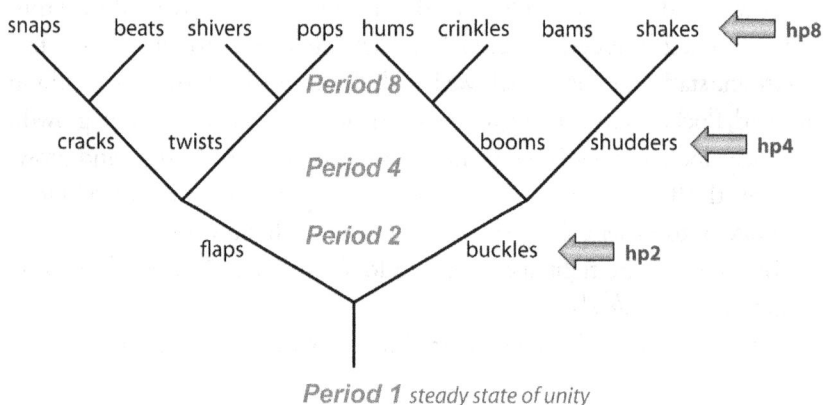

Periods 1, 2, 4 & 8

In this b-tree, the trunk at first forks into the 2 branches of *Period 2* that can be read across *horizontally* as the events in a *horizontal period 2 window (hp2).* Then it forks again into the 4 branches of *Period 4* that can be read as its *hp4.* It forks yet again into the 8 branches of *Period 8* that can be read as its *hp8.*

Thus each time the b-tree's branches bifurcate, they multiply by a factor of 2, so that each new level of growth doubles the previous period's number of branches. That's why bifurcation is also called *period-doubling.*

2. A horizontal period 3 window tracks patterned chaos

A b-tree can keep on forking upward until it is filled with many dark branches criss-crossing each other. In this next image, the b-tree has forked so often that it no longer shows the clean, black lines of countable branches. Instead, those sharp lines seem to have blended into blurry strata. The forking has doubled again and again, cascading until in the upper part of the graph, the layers of twigs all jam into a dark blur. To a casual eye, it's random chaos.

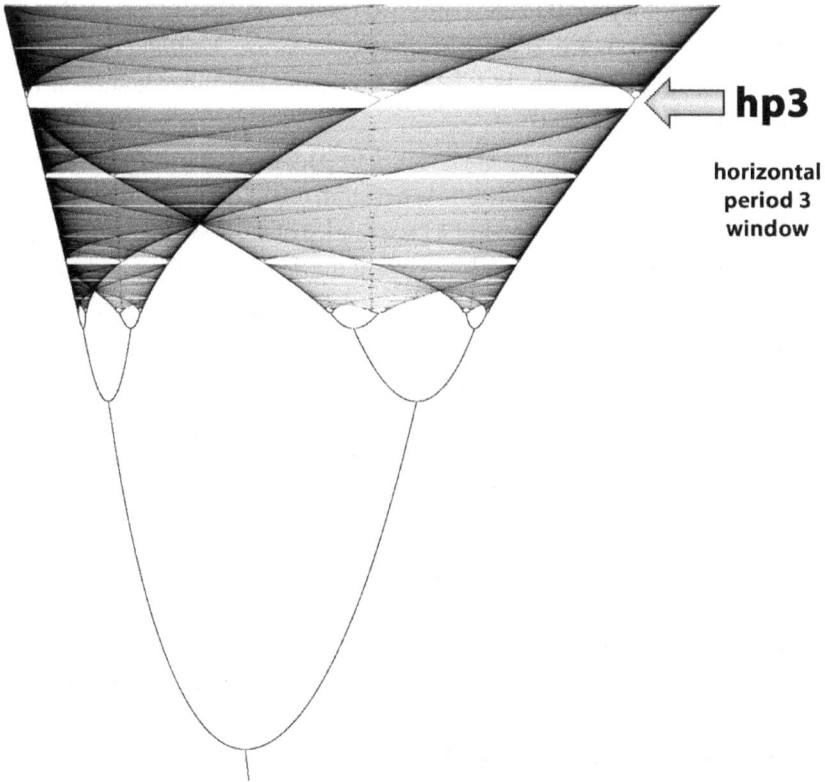

← **hp3**

horizontal
period 3
window

Bifurcation tree with a horizontal period 3 window (hp3)

However, not all of the branching is completely obscured in this b-tree. Notice how there's an occasional brief, clearer gap or window on some levels, where the branches fork less often? Occasionally a branch doesn't sprout, so sometimes a window may hold an odd number of branches. For instance, that arrow near the top of this b-tree points to the gap of a *horizontal period 3 window (hp3)*. Luckily! Why is it lucky? We'll investigate the significance of the hp3, for something grand comes of it.

If enough twigs ever do reduce down to just three branches in a gap, then above that hp3, the events are redefined. The branches above that hp3 are no longer tracking mere random chaos. Instead, they are now tracking *patterned chaos*. It may still *look* random, but it's not. Above that hp3, the branches hold recurring, deterministic patterns in their multiplicity of events. No matter how dense that twiggy welter, it will never again lapse into true random chaos. Instead, it is now fractal, *patterned* chaos.

James Yorke and Tien-Yien Li in 1975 proved that if a b-tree develops the hp3, then the branching events above it are no longer tracking mere random chaos. Instead, they are tracking patterned chaos. York and Li reported their findings in a famous article in *American Mathematical Monthly.* That article is packed with heavy math, but for our purposes here, the title sums it up: *Period Three Implies Chaos.*

There are other, more elaborate routes to patterned chaos. Yorke himself admitted that there's more to chaotic behavior than just the horizontal periodic window (hp3) in a b-tree. He called it "just one leg of the elephant, so to speak." Nevertheless, for us, the hp3 on a b-tree opens the door that will lead to *co-chaos*.

The hp3 in a b-tree assures us that its dynamic kernel will allow the bifurcations to drop back to that level, then sprout up again without losing any fractal mojo. The hp3 guarantees that the events it tracks can keep on proliferating in its branches, then drop back down to the hp3, then rise again in the branches to track random-seeming events that nevertheless keep iterating in a dynamic chaos pattern with variations. The hidden chaos pattern persists, although its details may shift continually.

Chaos systems churn out the way of the world. Emotional, social, cultural, economic, and political processes hold nonlinear chaos patterns, which means that we cannot identify the exact reasons for most events. Nor can we reliably predict what will happen. Still, we are assured by the hp3 that *something* will keep happening. This universe keeps going instead of dying like a bad battery.

The hp3 is important to chaos theory, and it is vital to the universe's master plan. Why? The hp3 with its 3-twig gap heralds the power of "matter with a will of its own," to quote Ilya Prigogine's phrase. A promise is embedded in the hp3. Instead of a point of no return, it becomes the point of all return.

You already realize that the thread of your life is not logically linear. Consider how it has pulled you into places, feelings, consequences that you never expected. Life defies your efforts to break down each event into merely linear, logical components that you can systematically analyze. It has tangled your personal life, your aims, your emotions into designs beyond any logical explanation. Reality does not docilely conform to your expectations.

Remember this: logic can chart various aspects of what has happened in a situation, and it can indicate some possibilities for the future, but rarely can it chart *all* of what has happened, much less what will happen. Far less, *why.* Indeed, if a *why* is offered, usually it turns out to be just a more detailed version of *what* happened *when, where, and how.* But never the overarching *why.*

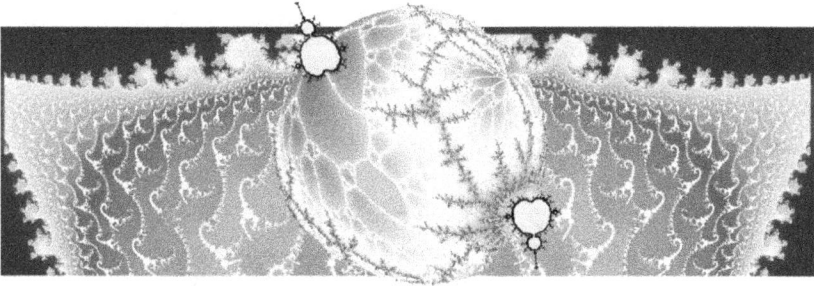

Chapter 10: Hearing Tings!

1. With such sharp logic, I cut myself

When God left me in that 1958 teenage dream, I felt the impact. When I woke up, I lost the message. My cynical, adolescent, logic-chopping attitude, honed on several years of debate tournaments, dismissed the flame-haired God so scornfully that I didn't even notice it was a *mystical* dream, for heaven's sake!

Here was the first-hand experience of God that I'd been seeking in all those dusty library shelves for years! But the dream came from within me, not from outside authorities whom I respected. And it held something I could not even acknowledge when I woke up…the loss behind God's words: "I am leaving now." Within the dream, those words scalded me…and yet, oh, the balm of God's touch on my arm as he promised, "But remember, I will come back."

In another dream in 1985, God finally did come back. That's why I'm now writing this book, this series…to describe in as much detail as I can muster what that return to the divine possibility brought me. It took me into the origin of space, time, matter, and energy; it showed me a master code that generated them. And most of all, I experienced the divine love behind it all.

As I experienced it, *saw/was* it, all the structural concepts of the universal architecture seemed to be just technicalities. I didn't know how to fathom any of it…except with love. I could not begin to identify or describe its physical layout, much less the intangible dynamics of space and time that I moved through, became, moved beyond. I could not encompass any of it in words, nor even in thoughts. I could only submit to that love's immensity.

Soft I might have seemed to that medical doctor talking with me about the *Tao of Chaos*. However, my path has not only blissed me in heaven; it also annealed me in hell. Looking back now, those awful years without God swung a far curve away on my road back. Those years of seemingly pointless, meaningless troubles insisting life is hard, and then you die. Until God came back.

The two dreams of God leaving and God returning were the two most important events of my life, yet they bypass waking logic. They cannot be measured or tested quantitatively. They cannot be verified in double-blind trials. They cannot be replicated in experiments using the scientific method, not in one dream lab, much less in many labs dotted around the globe. But they changed me, and I am glad.

2. Reading Dream Mail

Those sad-eyed, angst-ridden, European existentialists were wrong about me. As a teenager imprisoned in their view of an absurd world, I was not escaping the opiate of illusion; rather, I was dragged down by overwrought disillusion. The existential bitters that I sipped clouded my mind into viewing life as a binary series of *either/or* shunts tunneling blindly through reality's far vaster vista of analog connectivity. I couldn't sense the web of synchronicity, nor appreciate the innate meaning that embraces our universe and holds it together.

But thank goodness, for only so long could I endure my own unbearably sophisticated victory over clichéd hope. Back then, how naive it seemed of me to hope. Then God returned and showed to me...me, an educated woman, but a physics and math ignoramus...the amazing universal architecture that was created by numbers and love. It baffled my logical evaluation, yet it delivered my life into meaning beyond words.

That's why I mix physics with philosophy throughout this book, this series. It's the only way I can describe how the dream carried me into a place where a human body can never go, nor tools tap, nor words convey, into the mobic scale and below it to find universal mind housed in 3D time.

Down there, it orchestrates the evolution of our universe. It has slowly developed the upper bubble's conditions and made them fit to birth and nurture the many smaller bits of life sprinkled around up here. It gradually terraformed those accreting particles and meteors into our ball of Earth, burgeoned into life in its oceans and on its continents...into jazz and email and garbage dumps.

We can tap into the universal mind in 3D time with our own tiny minds. Indeed, we do it every night when we drop our egos and dream. In dreamtime, we tap into the universal knowing that transmits to us holographic messages uniquely tuned to each specific individual. That's even a bit how ants operate. Each tiny ant gets its own specific instructions on how to go about tasks, and together they form one greater intelligence that constructs the nest.

As your ego goes dormant in sleep, the larger universal mind offers you personalized dream mail that is coded by collective archetypes and personal symbols...whether your modern-day ego can read them upon waking or not.

The next morning, do you examine those letters delivered nightly in your dreams? Do you even notice them? Dreams are networking analogic, not linear logic. They speak a long-forgotten language still known to your soul, but as personal identification with the ego has taken over, it's now lost to consciousness.

Too bad we are so illiterate now that we cannot read our dream mail. However, the skill can be relearned. One of my joys over the past 35 years has been teaching dream interpretation to others, who will perhaps teach it to yet more. This soul mail appears in our dreams worldwide every night. It is delivered in dreams around the globe. We can all learn to read our Dream Mail.

3. Hearing voices...crazy or blessed?

Occasionally a message comes without a dream. I remember the first time I heard an inner voice...not in a dream, I mean. It was in 1986. My husband John and I had lived in the Kusnacht suburb of Zurich for close to a year by now. I attended Jung Institute classes by day and frequented the ETH library several nights a week, where I made notes on the genetic code.

At the Jung Institute, I was learning how to work with dream genres, interpret them symbolically, explore the issues they bring up, and so on. I was also experiencing Jungian dream analysis first-hand while completing at least 90 hours of analysis with male and female analysts.

By now, I'd clocked about 30 hours with Elisabeth Ruf, so it was time to switch over to a male analyst, Dirk Evers. I chose one, and we agreed by phone to meet and decide on whether we could work well together for a year.

Yes, "work together" is how Jungians describe the analytical process.

I took a suburban train into downtown Zurich, and then I sat waiting in a small reception area outside the analyst's office door. Could we find enough common ground to begin the process of analysis? Should I mention those two God dreams? After all, they were what brought me to the Jung Institute...but still, I felt nervous about mentioning it right away. Would he think me batty?

In the alcove, I heard a calm voice say, "Don't worry. I'll keep an eye on you."

Where did that voice come from? I even glanced around the alcove. Nobody! So...was the voice inside me? I panicked! Here I sat, waiting to meet a Jungian analyst, and *right now* I started hearing voices? Don't crazy people hear voices?

Then my mood turned wry. What a curious turn of events! How could I possibly tell this guy that now I'm suddenly hearing voices right in this alcove? He'd probably think I was nuts. I smiled at the irony of it. Such a twist!

Okay, so I'll just follow my motto: CKCQ. Courage. Kindness. Caution. Query. Silently I directed a thought at this inner voice: "Who are you?"

"God."

That threw me into an even greater panic. What! God? Oh no! Why this special monitoring of me right now? Was I about to flub this interview, no matter what I said or didn't say about God dreams or hearing voices? Silently I wailed, "But I don't want to be watched! Especially not by God!"

After a pause, the voice answered, "Then I'll go to sleep." It was gently spoken, but the implication nevertheless quelled me. So now I'm laying the law down to God? Saying don't pay attention to me? It threw me even more off-stride. I felt dithery inside, way out beyond my ordinary coping strategies.

Then the voice said slowly, kindly, with judicious deliberation, "Don't worry. I sleep with one eye open." And I received the distinct image of myself as a pearl resting in an enormous hand.

That nonplussed me. I thought the whole thing over. Here was somehow the best of both worlds. God would keep an eye on me, yet drowsily wouldn't peer at me too hard. I could hear God, yet God would not speak to me right now. And God treasured me, protected me like I was a valuable pearl. Overwhelming, yet somehow also comforting. All of this was a paradox greater than I could resolve. So I ceased to struggle and just pictured myself sitting there in the alcove like a pearl in the palm of a giant hand, God snoozing overhead with one eye open.

When the door opened, Dirk Evers invited me into his office. Soon he mentioned that he was a former Catholic priest. Within 15 minutes, I realized he knew the spiritual life and we could get along together even though I was not a standard Christian...well, maybe he wasn't, either. To my relief, before the hour was up, I told him about my God dreams...the first one in 1958, when God left me, and the second in 1985, when God came back. But not about the voice. I was still hesitant to admit that even to myself.

The guy immediately understood the importance of those dreams to me, and how much their hidden vortex had shaped my life. I felt an inner peace. This man could see why those two dreams had changed my life's direction without me having to spend a lot of time explaining it or justifying it to him.

After a few months of analytic sessions, Evers said he wanted to call the sessions a trade instead of me paying him for each hour. Why? Because he felt he was learning from me as much as I was learning from him. Okay, who was I to argue with that? Unexpected...but I accepted it. And by then, I'd decided that yes, it helped to hear a voice and get the sense of being a pearl in God's hand. The long-term benefits showed it wasn't necessarily crazy to hear a voice.

I began to feel that God, synchronicity, blessing, the grand organizing design, you name it, touched that first meeting with the Jungian ex-priest and turned it into a friendship. But can I prove to you that I heard a voice, and it blessed me? No. Can you prove otherwise? No. Nonetheless, it happened.

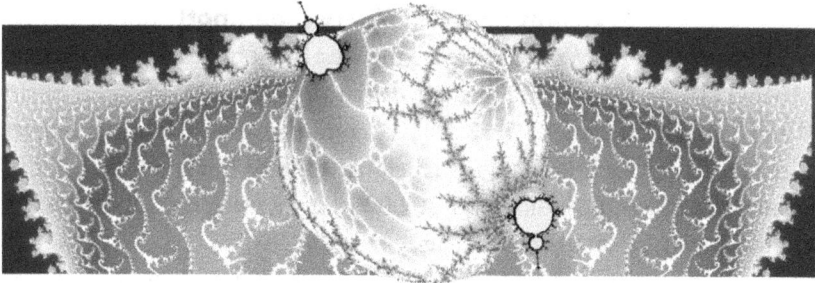

Chapter 11: Polarized Bifurcation

1. Polarized bifurcation

Co-chaos begins very simply. It starts with a bifurcation that is also polarized into *minus* and *plus*. The I Ching math symbolizes *minus* as yin ▬ ▬ and *plus* as yang ▬▬▬. How did the I Ching and its symbolism begin? To quote M. Alan Kazlev, "There are two histories of the I Ching, the mythological and the academic, and they are both sort of muddled." Whatever, wherever, whenever its origin, this mathematical shorthand grows on a *polarized* bifurcation tree. Thus it is not merely a *b-tree*, but instead, a *p-tree*. It starts as a gray seed of neutral unity. The seed grows a trunk forking upward into two polarized branches of black yin and white yang. The two poles are complementary, coexisting partners.

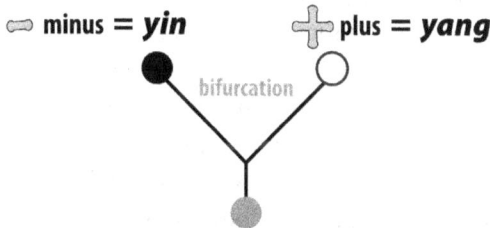

Bifurcation into 2 polarized branches

In Chinese thought, yin and yang have - and + polarity, so a Western-trained mindset might automatically equate yin and yang to the binary numbers of 0 and 1 that can shunt back and forth like an *off-on* switch. Leibniz did.

Binary off/on *switch*

2. Leibniz invented binary code...not!

Binary arithmetic was discovered in the West by Gottfried Wilhelm Leibniz (1646-1716), a German mathematician and philosopher. He used binary 0 and 1, the same symbols that a binary computer uses today. In Hanover, Germany, the Gottfried Wilhelm Leibniz Library holds about 100,000 manuscript sheets, including over 20,000 letters written by Leibniz and his approximately 1,300 correspondents.

In a manuscript dated March 15, 1679, Leibniz described his invention of binary numbers for calculation. That manuscript mentions "the possibility of designing a mechanical binary calculator that would use moving balls to represent binary digits." Thus by 1679, Leibniz had described binary arithmetic and touted using binary numbers in a mechanical binary calculator.

Leibniz worked for years on inventing binary machines that could execute arithmetic and algebraic operations. He believed that binary numbers could lead to new mathematical breakthroughs that were not possible with other number bases. But despite his best efforts, Leibniz never managed to build a completely realized binary computer, so his interest in binary numbers receded...

...until studying Chinese thought revived that interest almost 20 years later, when Leibniz started corresponding in 1697 with Joachim Bouvet, a Jesuit officially sent to Beijing by the pope. Bouvet's duty there was to teach the Kangxi emperor and his imperial court about Western mathematics and astronomy, and by that means, hopefully, build bridges to Christianity.

The Kangxi Emperor, the longest-ruling emperor of China—61 years— brought wisdom and stability during his reign. In an astute cross-cultural ploy, he introduced to Bouvet the I Ching by calling it an essential element of Chinese culture, declaring that Bouvet should acquaint himself with it.

Bouvet became so impressed by the I Ching math that in late 1698, he wrote to Leibniz a remark that yin and yang lines offered a very simple and natural way to describe some basic principles that exist in all the sciences. That slow-mail letter between the two men was pivotal in the history of binary numbers, for it stimulated Leibniz in early 1701 to write to Bouvet a reply that held a description of the principles of his own binary arithmetic.

Bouvet answered in late 1701 with a multi-part missive containing a Chinese woodcut of the I Ching's 64 hexagrams. They were laid out as a chart in both circular and square formats. After about a year and a half of circuitous, slow-mail misdelivery that even routed the packet through England, it finally reached Leibniz in Germany in 1703.

Following is an image of that woodcut. In its center, a square of hexagrams counts in binary from 0 to 63, reading from upper left to lower right. Around that square, an outer circle of hexagrams counts in binary using an elaborately bisected, mirror-image order.

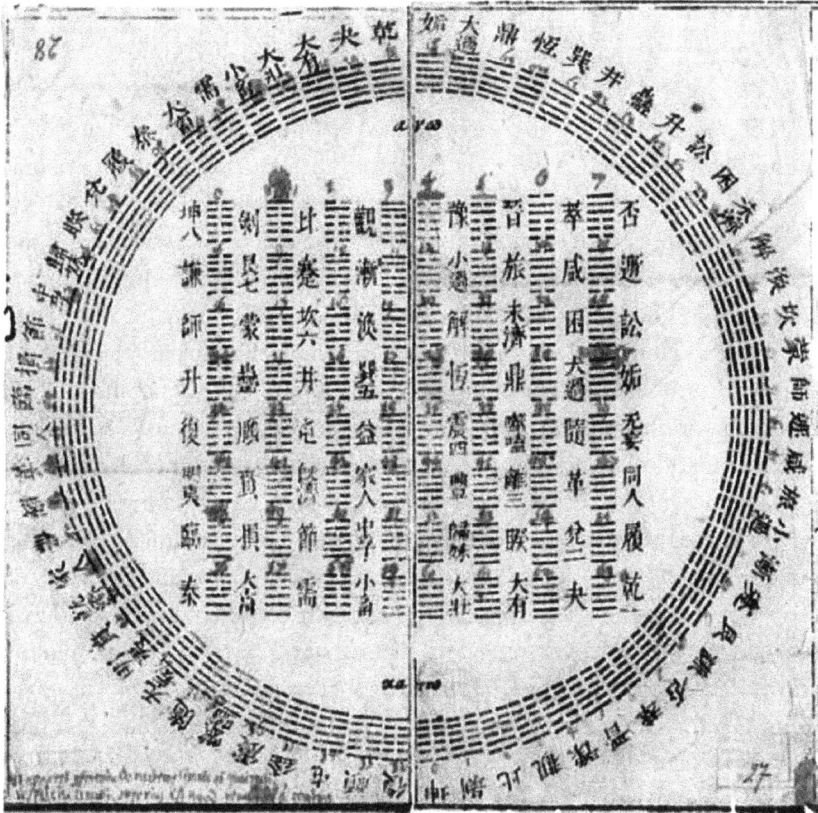

Circular/square chart of binary hexagrams sent by Bouvet to Leibniz in 1701

Leibniz Archive, Niedersuchische Landesbibliothek
(Look for the two corner numbers–27 & 28–added by Leibniz.)

Upon examining the I Ching chart, Leibniz realized with a shock that equating yin — — to 0 and yang —— to 1 would turn the 64 Chinese hexagrams into binary numbers counting from 0 through 63. What a shock! The remarkable binary arithmetic that Leibniz had just invented in Germany? So modern, so logical, so in tune with the dawning Western Enlightenment? Its cutting-edge math was already in the ancient I Ching!

Yes, Leibniz was right. Yin and yang can be tasked to symbolize the 0 and 1 of binary code. And yes, he was correct that binary code can be quite useful. Today human society throughout the world runs on computers, and most of them use binary code. Giant electronic brains compute using strings of 0s and 1s in governments, businesses, and homes around the globe.

I CHING		BINARY		DECIMAL
yin ● – –	=	0	=	*0*
yang ○ ———	=	1	=	*1*

Yin and yang as binary & decimal counting

This table shows yin and yang equated to binary 0 and 1 in the base-2 numerical system, and it also shows their number equivalents in the decimal or base-10 system. Sequencing in a binary string works like an *off-on* switch, forcing a choice at each step, shunting along as either 0s or 1s to develop the coding string. Only 0 and 1 are needed to initiate a long string of binary code—for instance: 1011100101111000110110010.

Perhaps Leibniz had assumed his mechanical binary calculator languished simply because the concept of binary numbers was just too far ahead of its time. But upon examining the 64 hexagrams that Bouvet sent to him, Leibniz realized binary counting already existed very far back in Chinese history.

Within a month, Leibniz wrote a paper in French comparing his binary system to the hexagrams of the ancient Chinese I Ching and mailed it off for publication in France. He also sent a brief account of it to England's secretary of the Royal Society. In it, Leibniz called binary arithmetic both a computational tool and a way to discover some deep mathematical, physical, philosophical, and even theological truths. He predicted that using base-2 numbers could bring together several crucial strands in humanity's world view regarding creation, order, and harmony.

Leibniz said of the I Ching's binary math: "...this arithmetic by 0 and 1 is found to contain the mystery of the lines of an ancient King and philosopher named Fuxi, who is believed to have lived more than 4000 years ago, and whom the Chinese regard as the founder of their empire and their sciences."

We can agree with Leibniz that the 64 hexagrams showing "this arithmetic by 0 and 1" count in binary from 0 through 63. Yes, that works, and it is definitely part of the story. The danger? If we see yin only as 0, and yang only as 1, such a mindset will miss the larger, nonlinear scope of I Ching math.

We Westerners may even equate yin as 0 to empty, passive, useless *nothing* and yang as 1 to weighty, assertive, useful *something*. That mindset can devalue yin into ineffectuality—"Yin's a real zero"—implying there's more value in acting like bold yang 1, a stand-up guy. Current buzz words (aggressive, pro-active, extraverted, can-do, loud and proud, in your face, kick-ass) tout and glorify yang's so-called "masculine" stance, as opposed to yin's wimpy, female, do-nothing 0.

Ironically, the 0 holds a mystery even in binary code. Claude Shannon, developer of information theory, saw this strange power in the nothing of 0 as a weird paradox. He said the nothing of 0 puts a peculiar kind of relational data into the coding string. In 1001010, for instance, empty 0s hold information by their very placement in the ongoing string. Shannon said the 0 holds "information by position" in a binary string. And digital number convention agrees. The 0 affects numbers through position—as in 10, 100, 1000—by calling each 0 a place-holder...as if 0 were merely a bookmark instead of part of the book. Paradox indeed! What and where is the true value of 0 here?

3. Seeing yin & yang in a "both/and" analog view

In Western thought, yang energy (assertive quantity of doing) is highly valued, while yin energy (receptive quality of being) is less valued. But ancient China appreciated yin as the dark mystery that receives, accepts, holds, and nurtures in silence. It is the flexible receptivity that may appear as outwardly passive and inert as soil itself, but it acts as the womb of creation. The Chinese metaphor for yin is dark, opaque earth. Soil accepts all that is put into it, hiding it so that you don't even know what's inside unless something new emerges. The earth takes seeds before they sprout, and it takes us after we die.

Do not fall prey to supposing that yin and yang equate only to 0 and 1. Yes, they can represent binary code's *either/or* shunt between the nothing of 0 and the something of 1, but they can also represent two equally powerful poles.

Yin as 0 has a place-holder property in binary mode that is different from its polarized property in analog mode, where yin is no longer *nothing*. Instead, in a tension between both poles, yin becomes an equal partner to yang. They support each other by their very existence. You cannot have one pole without the other, just as a battery needs both a positive pole and a negative pole to work.

Thus yin's negative − and yang's positive + do not equate to the "bad" and "good" of simplistic moralizing. Neither pole of a battery is bad, wrong, or powerless. Likewise, every human personality holds both yin and yang aspects, even if a woman dresses "all girly" and a guy acts "all macho"...or vice versa. Yin receptivity and yang assertiveness are both useful in an appropriate use. Moreover, either one may become too much, too little, or ineffective in unsuitable use. Then it's time for a change, a flip-flop, an enantiodromia.

Thus yin-yang polarity is inherent in the Tao's natural flow, and you cannot get rid of it by pretending that yin is only a zero, just as you can't erase one magnetic pole of the Earth by outlawing it. You can't even get rid of one pole on a magnet by cutting it off. If you cut off one pole of a magnet, you just wind up with two smaller magnets, each of them now sporting its own pair of poles.

Even the shapes of 0 and 1 suggest an amusingly anatomical female-male aspect. The 0 is rounded, maintaining a receptive space within its hollow center.

The 0 rings in nothing, yet it can be doubled into the infinity loop of ∞.

But look at 1! Sharply erect, it stands upright in linear assertion. Actively vertical, the 1 declares its proud, forceful existence in a bold stance of "I am!"

4. Three levels of polarized bifurcation on the p-tree

Yin and yang poles appear on a p-tree at the first level of polarized forking.

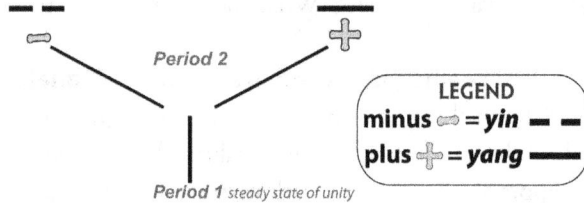

1 level of polarized development gives yin & yang

At the second level of forking, four bigrams offer a more nuanced polarity. (*Always read I Ching math from the bottom up, for that's how polarity grows.*)

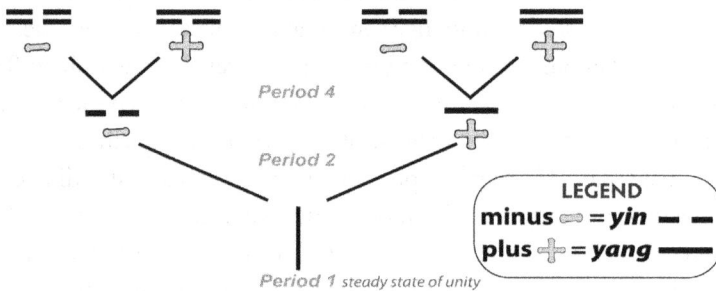

2 levels of polarized development gives 4 bigrams

The four bigrams are stable yin ⚏, stable yang ⚌, changing yin ⚎, and changing yang ⚍. A stable bigram has the same polarity on both levels of forking on the p-tree, but a changing bigram turns into its opposite pole at the second, higher level of forking.

When a bigram is read as binary numbers, it is like a stack of unitized bricks. Stable yang ⚌ is two consecutive 1s. Changing yang ⚍ is a stack where the sequence switches from 1 to 0. Here are their binary and decimal equivalents:

BIGRAMS		BINARY		DECIMAL
⚏	=	00	=	*0*
⚎	=	01	=	*1*
⚍	=	10	=	*2*
⚌	=	11	=	*3*

Bigrams as binary and decimal counting

However, there is also a polarized, relational, analog way to read bigrams. It sees a bigram as two poles or attractor points, not just two units. In this view, rising from the first fork's simple yin/yang polarity, the second fork's more complex polarity has four attractor points.

As a metaphor, by developing bigrams, the I Ching shorthand has jumped beyond black and white to include two shades of gray. At this higher level of perception, things no longer look just *black* vs. *white, no* vs. *yes, bad* vs. *good,* villain vs. hero. We've now begun to realize that there's some good in the bad and some bad in the good. That's why the tai chi symbol (☯) nestles black yin and white yang like a pair of fishes, head to tail, with a dot of each inside the other.

At the third level of forking, the four bigrams get more nuanced into eight trigrams atop the p-tree. Metaphorically, we've moved past grays into colors.

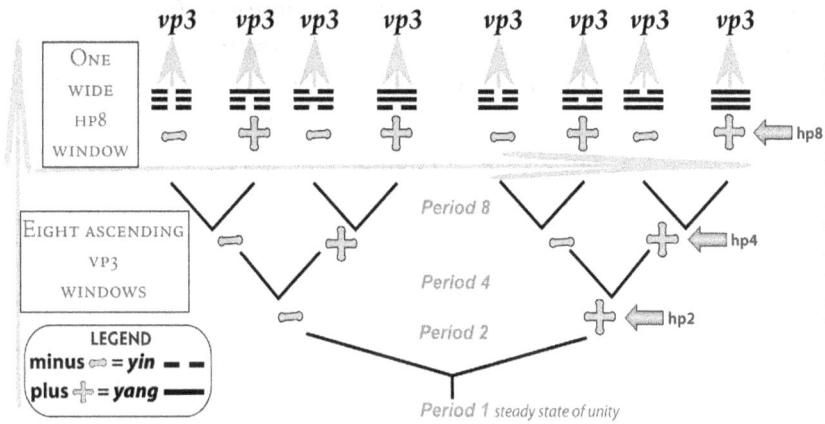

3 levels of polarized development gives 8 trigrams

If you read this p-tree's third level *horizontally*, you find a horizontal period 8 window (*hp8*). But if you read this p-tree *vertically*, you find 8 different vertical period 3 windows (*vp3s*). Each trigram is a vp3 shorthanding its own unique chaos pattern. By forking in just three levels of branching, the p-tree has grown eight different trigrams of chaos patterning, yet they all share a family identity.

All eight trigrams make an octave of chaos patterns. They combine binary units and analog polarity in a remarkable way, with a number makeup so very fail-safe and self-reinforcing that I call it more than nonlinear. It is analinear.

And since by just the third level of forking, a p-tree develops eight vp3s of chaos power described by the eight trigrams, the Chinese did not bother to keep adding more lines atop each trigram. No need for a fourth, fifth, or tenth level of branching in the p-tree. Certainly no need for that welter of tiny twigs in an ordinary b-tree, which despite its many forkings, can only offer horizontal period 3 windows (hp3s) of chaos patterning.

5. The double p-tree grows branches and roots

All 8 trigrams in the family octave produce a nuanced harmony that can also become part of a yet larger opus. What larger symphony, and how?

No more upward branching for the p-tree! The ancients did something quite ingenious to bring more reach into the system and give it a still-vaster order of complexity. They balanced the p-tree's branches with a parallel set of roots forking down through three levels. Thus it now becomes a double p-tree—or for short, a *dp-tree*. Its two different yet related domains hold 2 octaves of 8 trigrams each, and each trigram is a chaos pattern. The trigrams can pair-bond across the whole system in 8 × 8 possible combinations to make 64 hexagrams of *complementary chaos*—or for short, *co-chaos*.

A hexagram has 6 lines…from Greek *hex* meaning *six,* and *gram* meaning *line.* Even the Star of David is a kind of hexagram since it has 6 lines. Just as the Star of David is not only 6 lines but also a pair of interactive triangles, likewise, the I Ching hexagram is not only 6 lines but also a pair of interactive trigrams.

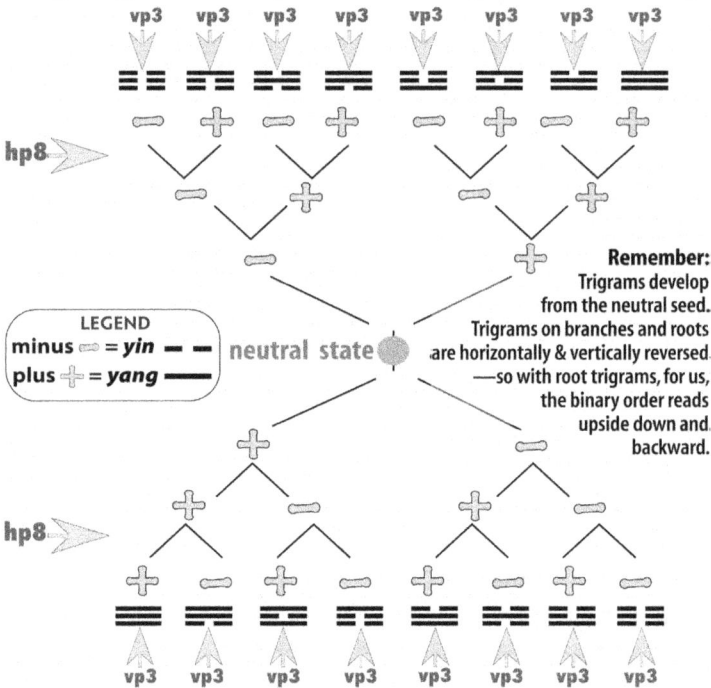

The double p-tree generates hexagrams

In a hexagram, each of its two trigrams is a chaos pattern described by its vp3. Thus a hexagram contains two vp3s that bond into a co-chaos pattern. The bonded pair of trigrams may interact constructively or destructively, depending on their wave interference…and recall, sometimes destruction is a good thing.

Together, all 64 hexagrams represent a vast polarized system of 64 different co-chaos dynamics spreading across the roots and branches of the dp-tree. This new, larger co-chaos system has a dynamic complexity that is far more comprehensive, versatile, and efficient than mere chaos patterning, yet every aspect of it stays fully associated in all its levels and polarities. Everything relates.

When the cosmegg's nothing discovered itself as something, the number 0, it could switch from nonbeing to being by pulsing *off* and *on*, so that now 0 or 1 can represent the *off/on* of nonbeing or being in binary mode. Moreover, in analog mode, 0 also equals –1 and +1.

In each option, 0 and 1 operate. Call them yin and yang, a polarized pair that evolved in each domain of the dp-tree into a polarized pair of pairs: 4 bigrams…then into an octave of polarized triplets: 8 trigrams…as chaos patterns…bonding into 8 × 8 pairs of polarized trigrams = 64 hexagrams. These 64 dynamic co-chaos patterns also count in binary! Here is the paradigm.

Now, to apply the paradigm…polarized pulsing at the mobic scale established triplets in spacing and timing that I call griplets. In the hourglass cells, they bonded into 8 × 8 pairs = 64 possible grips that projected and still hold together our Double Bubble universe's space-time lattice. Its analinear math merges the linear *either/or* style of binary numbers with the analog *both/and* style of period-doubling/exponential growth to develop 64 co-chaos patterns of polarized force, funneling all those number approaches into the transcendent condition that jump-started our living universe from nothing.

A latter-day variant on that paradigm is our own DNA. Triplets called codons bonded into 8 × 8 pairs = 64 DNA swatches that hold together the double helix. Other lesser variants (often with broken symmetries) include octaves of notes in music, octaves of chemical elements in molecules, octaves of quarks in hadrons, and octaves of waves in the electromagnetic spectrum.

And the I Ching math can shorthand all this! The dp-tree's reversing mirror of bifurcating branches above echoes its plunging roots below. Somehow the ancient Chinese found the same *ur*-order that I saw in a dream as candelabra reflected on a mirrored table. I saw it/was it. So are we all, for our DNA carries the 8 × 8 pairs = 64 different basic 6-packs of co-chaos patterning.

6. An extra for the math-minded…

If a p-tree's roots just kept on forking, this could be shown as binomial expansion in a triangular shape. It was done in ancient China, Persia, India, Greece, Germany, and Italy. Records in China say that Jia Xian developed this triangle in 1050 CE, but its earliest extant image is by Yang Hui (1238–1298 CE). His triangle allows positive integers to show how hexagrams employ binomial expansions that can mathematically describe exponential growth.

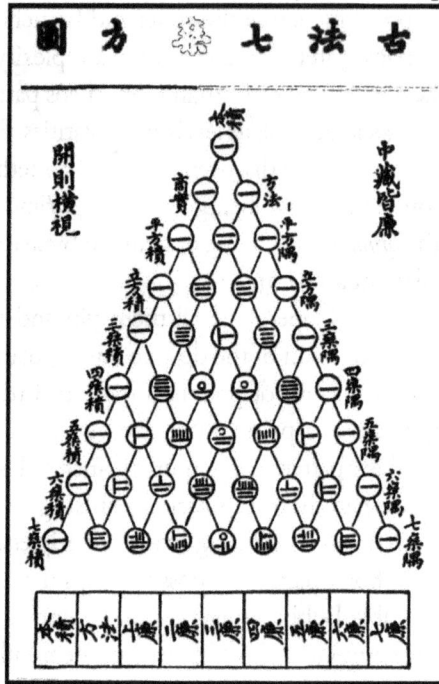

Yang Hui's Triangle

The same triangular array of binomial coefficients can be seen 600 years later in the Pascal Triangle below, which shows only Rows 0 through 6 of Yang Hu's triangle. The odd numbers are light; the even numbers are dark.

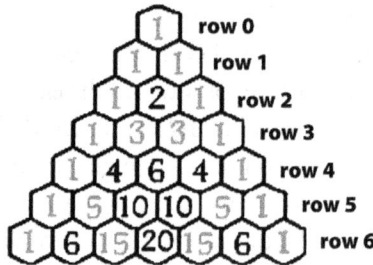

An abbreviated Pascal's Triangle

All of the outer *1s* form a boundary on both sides. Inside that boundary, each dark number is the sum of the two numbers just above it. (This shows up better in the colored ebook version.)

The same math process also develops the Sierpinski gasket or sieve, shown below. *Stage A* colors in the areas of odd and even numbers in rows 0 through 6 of Pascal's Triangle and develops three consecutive tessellated triangles. Light triangles symbolize odd numbers; dark triangles symbolize even numbers.

A

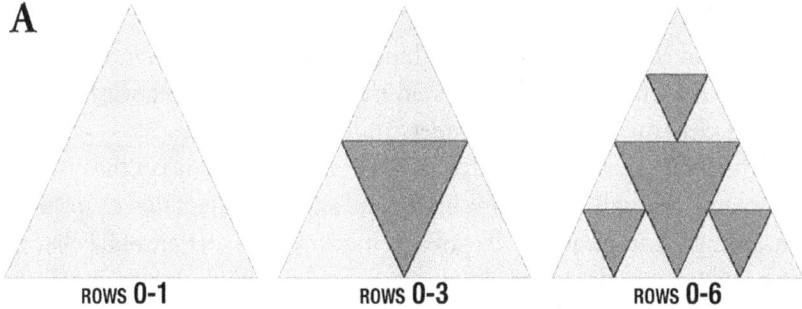

ROWS **0-1**   ROWS **0-3**   ROWS **0-6**

B

C

Sierpinski sieve progressions

Stage B of Pascal's Triangle develops more rows of smaller pyramids stacked within the same 2D space to establish an increased fractal complexity. This higher intricacy inside a fixed boundary at each new iteration is characteristic of fractals. Wacław Sierpinski described this mathematical development in 1915, but its motif had appeared in Italian art by the 13th century.

In *Stage C*, for fun, you can see a 3D, square-based version called a Sierpinski sieve, along with its theoretical "inverse."

Other cultures also developed variant expressions of this mathematical patterning, and they all use the same, short binomial expansion: $(x + y)^n$.

7. The effortless work of Wu Wei

Co-chaos merges the sequential *either/or* of binary numbers with the analog *both/and* of period-doubling and exponential growth to generate you and me. Remarkably, a brain does something similar when it generates thoughts. It reconciles left-brain and right-brain modes into the higher synthesis of a transcendent third output: the understanding of something.

That's why I suspect that artificial intelligence cannot successfully imitate a human brain without merging binary and analog computation in analinear processing...and paradoxically gives us humans our deterministic behavior, along with a measure of free will, beware of the result if it appears in AI!

In 1987, I was walking a footpath alone in the high, green meadows of the Swiss Alps. After some time, I came to a cottage with the words *Wu Wei* painted on its front wooden gate. I knew the phrase was Chinese...but why was it painted here? I stood at the gate, listening to music that came from...where?

Then I saw a young Swiss man working in a goat barn as Vivaldi's music cascaded from a device on a shelf behind him. I'd never before come across a goat barn rigged for music. When the man saw me, he came over to talk.

I asked him why *Wu Wei* was painted on his gate. Quite the world citizen, like so many Swiss, he said the gate was named *Wu Wei* because it means "doing without overt action." Just by sitting there, he said, the gate keeps his herd of goats inside the fence. He went on to say that it was his dream to live in *wu wei* with his extravagantly beautiful environment instead of wrestling it into submission. He wanted to hold his land by being in tune with it.

In the co-chaos paradigm, binary numbers insist on a win-lose standoff of *either/or*, but that is sweetened by the *both/and* of yin and yang's relational, analog coexistence. Together, binary and analog processes achieve the strange possibility of walking several paths at once—so that simple 2, whether it is moving as $2 + 2 + 2 + 2 = 8$, or going as 2 doubled to 4 and redoubled to 8, or exponentially as 2^1, 2^2, 2^3 going to 8, or by a mix-and-match routing—it all still leads to the same goal of 8. This unique ability of the 2 to tolerate and utilize so many routes to 8 is what melds the paradigm into analinear synthesis. Chapters 13 and 17 contain more about the 2's unique gift.

Co-chaos provides the paradigm for life itself. It is honored in the Hindu concept that says the masculine principle is surrendered assertion, while the feminine principle is active receptivity, and together the two principles generate birth. Co-chaos holds the art of doing much with little...just some 2s. Its power is subtle but effective. It hides at the heart of the I Ching's yin-yang shorthand of the polarized pulsing that generated our Double Bubble universe.

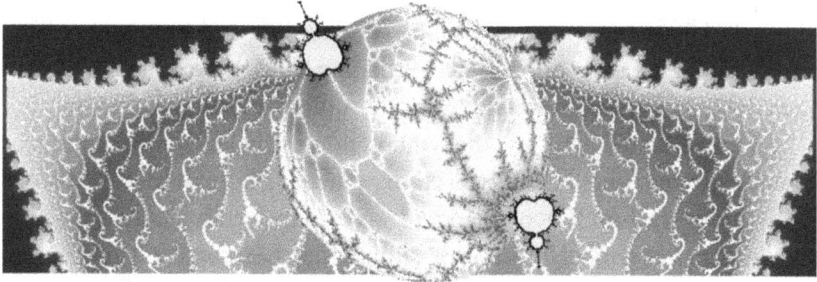

Chapter 12: Hearing Tings!

1. I make polarizing Mobius bands

One snowy weekend during the winter of 1987, the doorbell rang at our apartment in Switzerland. When I opened the door, a German friend handed me a book. He said, "Katya, I bought this book yesterday, but when I started reading it, I decided I must have picked it up for you."

He gave me Martin Gardner's *The Ambidextrous Universe*. That gift book helped me verbalize the dynamic I'd seen in a dream but couldn't explain. I'd seen a mobic dynamic in the membrane interface where space and time emerge, far tinier than the quantum scale where matter and energy emerge.

On that gray Sunday afternoon, I sat at my desk reading the gift book. My husband sat on the couch reading a Len Deighton spy novel. Sometimes I'd pause and look out the big, thermal-paned window to our snow-covered yard and the black-iced street beyond. By the curb stood a graceful cherry tree. White snow was banked on its bare, dark branches that in summer gave sweet, red fruit.

Turning back to Gardner's book, I read the next chapter. It suggested that I make a Mobius band from a paper strip. Shrug. I did so. Next step: Draw a sawtooth line along the band, zigzagging from edge to edge on both "sides" of the band through all 720° of rotation. Shrug. I did so.

Then I watched the sawtooth mountains flip-flop as I rotated the twisty band. They'd alternate between standing on their bases and their points. Something was ringing in my head like a crystal chime, coaxing me to pay attention: "*Ting!*" I heard it, but not with my ears. Attention to what?

So I made more Mobius bands. For over an hour, I sat at my desk cutting strips of paper, making Mobius bands, decorating them with symbols, twirling them in my fingers, glancing out occasionally at the black silhouette of the cherry tree at the curb, wondering what all this pondering was in aid of?

When I eventually tired of cutting paper strips, I began to rip margin strips from sheets of pinhole-feed paper (common in home printers back then). I'd take a thin strip, one edge rough from the rip, and with holes punched along its length. I'd half-twist one end of the strip and then tape it to the other end.

Voila! I had a hole-punched Mobius band turning under my fingers. That rough, torn edge along one side of the rotating band showed me how,ß with each 360°, its torn edge reversed from left, to right, to left...on and on.

So then I started drawing little symbols near the holes. Simple things. For instance, on one band, I drew small crosses extending from the punched holes. Then during the first 360° of rotation, I watched each cross and hole together form the ancient symbol for Venus ♀ . But during the second 360° of rotation, Venus would flip upside down, instead becoming the old symbol for Earth ♂. Twirling the mobic band flipped their two identities back and forth.

To my fascinated eye, they were transforming into each other right before me...easily, endlessly along the twisty 720° path of the band. Hey! If ♂ stood for Mars instead of Earth, then they'd be doing polarity switching. The old Roman symbol for *feminine* Venus would turn into *masculine* Mars! Hmm. Just as *yin* and *yang* transform into each other on the tai chi circle. And like the sawtooth mountains zigzagging along the Mobius band, repolarizing to stand on their bases or stand on their points. Their flip-flop did polarity switching.

Meanwhile, a crystalline *Ting! Ting! Ting!* in my head kept insisting, "Something else goes on this polarizing path, something important." But what? I wasn't yet able to correlate the evident polarity-switching on a Mobius band with what I saw in that great dream. I didn't recognize it yet, but I was holding a parallel to what I encountered there, clarifying how pulsing in the membrane's polarizing mobic pores can generate space and time for the two huge bubbles.

It was not in any of the physics books I'd been reading. Sure, current physics recognizes the tiny quantum scale where particle-waves emerge, but it does not yet recognize the far-tinier mobic scale where space and time emerge.

Twirling the bands, I thought about how DNA sets its polarized codons into 6-packs along both spirals of the double helix. I wondered...what if those two spirals functioned like a variant of the polarizing sides of a Mobius band? What if, instead of spiraling in a double helix, codons somehow sat along the loop of a Mobius band? How might such a code operate along that infinite yet closed path of ∞? What sort of code would even allow such activity?

Watching the paper bands turn in my fingers, I felt suspended, stymied because I didn't see what awaited me ahead—I only knew that on the horizon of time, something large was looming. Or ultra-tiny? Why was I sitting in idle play for two hours of self-induced trance over some twisty bands? I didn't yet recognize that I'd found a parallel to the peculiar polarity-switching along the endless 2DD surface of each mactor at the mobic scale. Not consciously, at least.

Nor did I grasp that I was finding a way to describe the master code running along each mobic pore of the interface that generated space and time for the two huge bubbles of our universe. Nor had I yet read that physics was seeking a way to describe the emergence of space, time, and gravitation via information. I was just trying to understand the dynamics of what I'd seen with the wordless naiveté of a baby in that God dream. Playing with Mobius bands helped coax forth the clues slowly coming together enough to let me propose this TOE.

2. The tings make a trail of breadcrumbs

For 5 years in Zurich, I worked on correlating the math of the 64 hexagrams with the 64 DNA codon 6-packs that run along the double helix. As I worked, clues kept appearing in my dreams like breadcrumbs. Even awake, my inner eye somehow kept visualizing four dancing bricks etched with yin and yang symbols. Their lines were polarized pairs that danced a squared-off ballet on some strange stage. But what stage? Those dancing bricks made no sense to me.

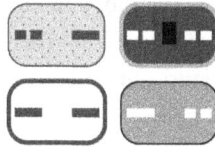

Four polarized bricks can dance

But I eventually learned at least enough physics to glean the significance of what that choreography showed me: how polarity-switching works down at that ultra-tiny scale. Its membrane interface is made of tiny mobic bands whose twisty surfaces are polarized by 2D space and 2D time. When I realized that each twisty pore combines some traits of a Mobius band and a Lorenz attractor, I began to call them *mactors*. Polarized pulsing drums along those mactors to generate space and time, plus the potential for matter and energy, projecting it all outward beyond the mobic scale as two holographic, mirror-twin bubbles. That pulsing still flings upward our constantly refreshing spacetime bubble holding lots of mini-minds. It also still flings downward the timespace bubble holding one great universal mind.

And now I can finally admit aloud that even from childhood, I've heard an occasional crystalline *Ting!* trying to get my attention in some liminal zone at the edge of consciousness. Always it makes me feel like a wild animal in the woods that looks up without knowing why. Its *ting* always directs my attention to something deeper, wiser, stronger than my own sphere of awareness.

Eventually, from a gradual willingness to attend and follow the way of the Tao came the 10,000 *tings*—allow me a Taoist pun here—that fill this TOE. I now recognize that even from childhood, whether I realized it or not, those inexplicable moments of synchronicity nudged me onward—faster, easier, as I began to pay attention—to this moment when you and I meet in this account.

3. A knowing beyond consciousness

For centuries now, Western science has fixated on the narrow chutes of *either/or* logic that makes forced choices—as if our minds must emulate binary computers, or at least pay ideological lip service to the linear rules that govern Newtonian mechanics. We've lost our connection to a deeper, nonlinear timing embedded in organic reality. Today, many people view life's events as random chance. Some even envision a tinny techno-God above a sterile matrix that they claim we must try to escape. Others say we're a game on an alien computer.

But by moving into the liminal state beyond ordinary consciousness, one can perceive that we live in a living thing...our universe. In that larger perspective, we can start to perceive what awaits at the edge of time to be born...ideas, events, even people. For instance, I somehow met my granddaughter Sylvia before even knowing that my daughter was pregnant.

By then, I'd left Switzerland and was teaching at Jinan University in Guangzhou, China. Why? It allowed me to study the I Ching in the home of scholar Zhang Luanling. One night, a dream prodded me to go shopping downtown the next day. I didn't know why or what for. Without rhyme or reason, I climbed uneven, wooden steps in an unfamiliar store to its fourth floor. I was in an infant's clothing section that I'd not sought. I found myself buying a tiny pink silk brocade jacket. But why?

Back in my apartment, instead of folding that beautiful pink jacket away, I hooked its little hanger over the closet door handle. I kept staring at it, fascinated, wondering, "Why was I impelled to buy this pink coat? Why does it fascinate me so? Why do I consider it so beautiful? What a goofy thing to do!"

It hung there for 6 weeks, and then my son phoned from Texas to say that my daughter in Pennsylvania had just found out she was pregnant. Oh! So that's why I was celebrating I-knew-not-what with something so tiny and pink and exquisitely made. I sent it on to Sylvia when she was born.

That is just one step on the winding path I've followed while writing this series. Logically speaking, how silly and naive it was of me to walk blindly toward a way of embracing two kinds of physics, meta and physical...especially since I began this journey with no idea where I was going, nor what concepts I was seeking, nor how to deliver them to you. So you're welcome to view this account as a long, waking dream if you choose. That may be quite right.

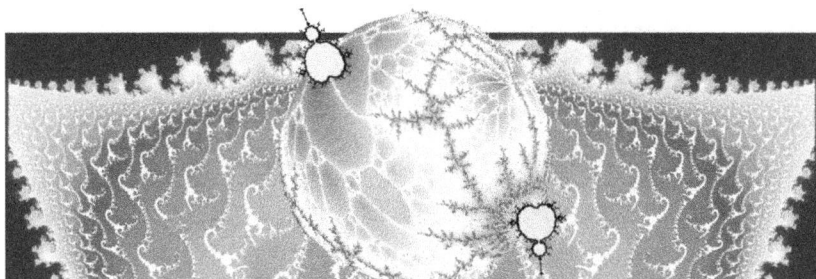

Chapter 13: Number's Secret Shuffle

1. What is a trigram made of?

To make sure the analinear number function that operates in this paradigm is truly clear, explain this another way. The polarizing, multi-level process on this chart below generates the simple math of Chinese trigrams, DNA, and dimensionality. Its bottom gray layer represents a neutral state, like the 0-number seed that grows a *polarized bifurcation tree* (p-tree).

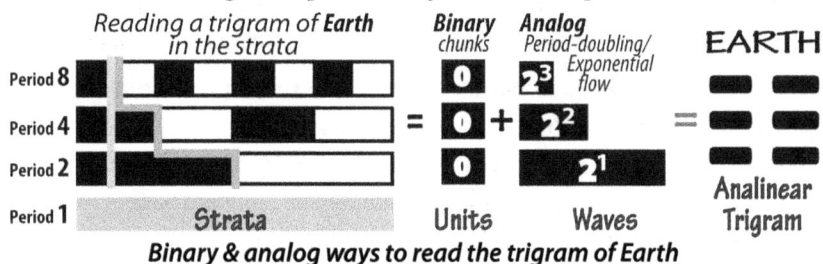

	Reading a trigram of **Earth** in the strata	Binary chunks	Analog Period-doubling/ Exponential flow	EARTH
Period 8		**0**	2^3	
Period 4	=	**0** +	2^2	=
Period 2		**0**	2^1	
Period 1	Strata	Units	Waves	Analinear Trigram

Binary & analog ways to read the trigram of Earth

Each new, higher level of bifurcation doubles its number of segments. This process is called period-doubling, and it is common in nature. It corresponds to change behavior at the onset of fractal chaos patterning. "Period-doubling occurs in fluid convection, water waves, biology, electricity, acoustics, chemistry, and optics, to name a few…" says Garnett P. Williams in *Chaos Theory Tamed*.

But note, in this TOE's variation on simple bifurcation, the rising segments on the chart, or on a p-tree, are also polarized into *yin* (black) and *yang* (white).

The bifurcating layers of strata can be chunked into binary numbers. For instance, run a vertical line (gold in the ebook) down through the strata to get a stack of three black bricks on the left end. They are unitized chunks, uniform in size, weight, value, significance. A black brick = 0. A white brick = 1. Thus the three black bricks on the left end = 000. The ancient Chinese called this the all-yin trigram of *Earth*, and they wrote it like this: ☷ .

However, you can instead do an analog, relational reading of the very same *Earth* ☷ trigram. Now you no longer view its layered, polarized segments as equal chunks. Instead, you treat them like electrical waves; each higher level is shorter, and its frequency and power will double/increase exponentially.

As another example, below you see the reading of a different trigram in the strata: *Fire* ☲. This trigram holds both black *yin* and white *yang* segments.

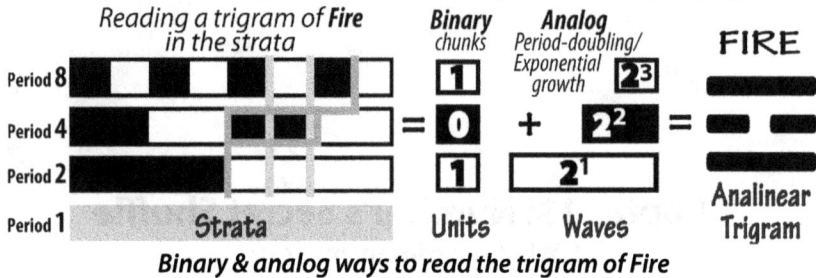

Binary & analog ways to read the trigram of Fire

For a binary reading of the *Fire* ☲ trigram, run two vertical lines (gold) down through all three layers of the black-and-white strata to get a stack of three bricks. White, black, white = 101.

But for an analog reading, use the stairstep lines (green in the ebook). They view the trigram of *Fire* ☲ as changes in wavelength and power. This approach yields two analog readings (doubling and/or exponential growth).

2. Trigrams use binary counting of unitized chunks

To understand the combo clout of numbers hiding in the trigrams, first consider some binary aspects. Below, you see all 8 trigrams shown as stacks of colored bricks. You also see their binary and digital equivalents. This sequence of trigrams automatically counts in binary from 0 to 7.

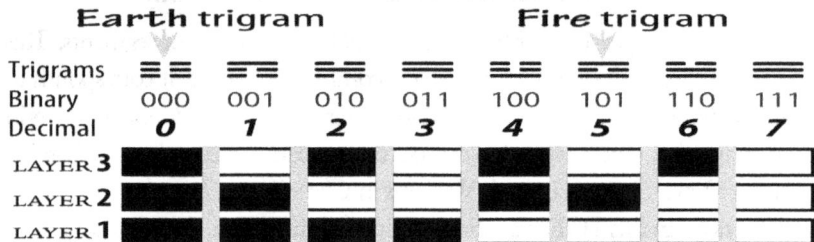

Trigrams								
Binary	000	001	010	011	100	101	110	111
Decimal	*0*	*1*	*2*	*3*	*4*	*5*	*6*	*7*

The p-tree order of 8 trigrams as binary & decimal units

This graphic equates trigrams to the third level of binary, *off/on* switching.

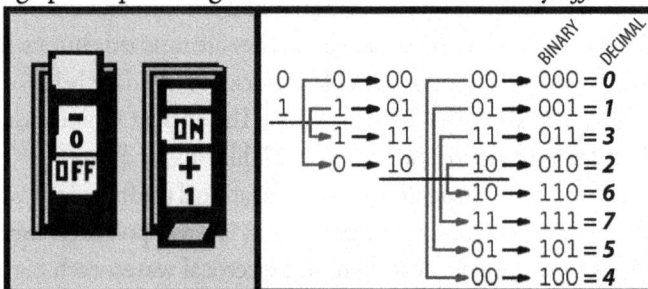

3 levels of binary off/on **switching**

To read a trigram as a binary number, mentally turn it so that its top line sits to the right. For instance, the trigram of ||| equates to 000, while the trigram of ||| equates to 001, and ||| equates to 010, and so on across the row.

3. Trigrams use two methods of analog flow

Next, let's turn to an analog, relational view of the all-yin *Earth* trigram. It sees the lines as waves of vibratory energy rising up through three levels. Each higher, shorter wavelength vibrates twice as fast as the one just below, so its energy doubles/grows exponentially at each new level.

This chart shows how both analog options echo the same number path from 2 to 8, thus reinforcing a surety of analog flow.

Earth ☰ ☰ has two analog readings =

	Period-doubling	Exponential Growth
	$2 \times 2 \times 2 = 8$	$2^3 = 8$
=	$2 \times 2 = 4$	$2^2 = 4$
	$2 = 2$	$2^1 = 2$
Strata	Analog reading	Analog reading

Bifurcation generates analog period-doubling in octaves of music

By viewing a trigram as period-doubling, you can compare its lines to octaves in music. In the trigram of *Earth* ☰ ☰ , its bottom yin line could compare to a long string vibrating out the deep note of low, low *A*. But if you fret it at ½ string, you can play low *A*, vibrating an octave higher and twice as fast. And by fretting it at ¼ string, now middle *A* vibrates another octave higher and again twice as fast. So *Earth* ☰ ☰ is *sort of* like three octaves of *A*.

EARTH trigram as period doubling	3 octaves of A	**Note**	**Frequency**
	1/4	middle A	$f = 440\,\text{Hz}$
	1/2	low A	$f = 220\,\text{Hz}$
		low, low A	$f = 110\,\text{Hz}$

Bifurcation generates analog period-doubling in octaves of music

Nature's octaves have long fascinated scientists and mathematicians. John Newlands developed a *Law of Octaves* for the elements in 1865, predating Dmitri Mendeleev's *Periodic Table of the Elements* by 4 years.

Newlands described his work this way: "Members [in my periodic chart] stand to each other in the same relation as the extremities of one or more octaves of music. Thus in the nitrogen group, phosphorus is the seventh element after nitrogen and arsenic is the fourteenth element after phosphorus, as is antimony after arsenic. This peculiar relationship I propose to call *The Law of Octaves*."

The analog reading of a trigram as exponential growth might even view it as a triple-decker fraction! For instance, in *Fire* ☲ , the bottom line is yang; the middle line is yin^2; the top line is $yang^3$…in three levels of exponential increase.

P-tree	Trigram	Binary		Doubling		Exponential
	EARTH					
Period 8	▬▬ ▬▬	0		doubly doubled yin		yin^3
Period 4	▬▬ ▬▬	0	&	doubled yin	&	yin^2
Period 2	▬▬ ▬▬	0		yin		yin
	FIRE					
Period 8	▬▬▬▬▬	1		doubly doubled yang		$yang^3$
Period 4	▬▬ ▬▬	0	&	doubled yin	&	yin^2
Period 2	▬▬▬▬▬	1		yang		yang
	HEAVEN					
Period 8	▬▬▬▬▬	1		doubly doubled yang		$yang^3$
Period 4	▬▬▬▬▬	1	&	doubled yang	&	$yang^2$
Period 2	▬▬▬▬▬	1		yang		yang

Viewing 3 trigrams as triple-decker fractions

There is something remarkable about this number path from 2 to 8. The 2 can take a chunky, unitized path to reach 8: **adding itself** (2+2+2+2 = 8). The 2 can also take two different analog, relational paths to reach 8: **doubling itself** (2 doubled = 4; 4 doubled = 8)…or performing the **exponential growth of itself** ($2^2 = 4$; $2^3 = 8$). Of course, the additive route makes an extra stop at 6, its hiccup spot circled on the chart below. Still, all three methods reliably hit 2 and 4 to reach the goal of 8. No other number path has three alternate routes that boast such a tight and total accord of method, progression, and goal.

The paths of **2** reinforce each other

2-Adding	**2**-Doubling	**2**-Exponentials
2	2	$2^1 = 2$
4	4	$2^2 = 4$
⑥		
8	8	$2^3 = 8$

Comparative paths of 2

4. Analinear math's secret shuffle: unitized chunks + analog flow

A physicist once told me he'd always just taken it for granted that the 2 can switch-hit its routes on the way to 8…that surely it was just a coincidence how adding 2, doubling 2, and the exponential growth of 2 all echo each other's stops to arrive at the sweetly coterminous goal of 8.

But no, the 2's ability to take three different yet reinforcing paths to reach 8 establishes a mathematical merger that is nearly magical. Each trigram describes a unique fractal chaos pattern whose number surety is secured by how those three internal paths of 2 moving to 4 and then to 8 will reinforce each other's accuracy. Its combo of binary counting/addition/doubling/exponential growth creates a tight, nonlinear result so special that I call it analinear. When all 8 trigrams appear on the p-tree, they offer 8 different, sequential vertical period 3 windows (vp3s) that can actually count off in binary numbers!

The 2's merger of method, progression, and goal between 2 and 8 is what lets our DNA remain stable as it paradoxically also evolves. That number accord is also what keeps our universe stable as it evolves. This series shows how the paradigm shorthanded by I Ching math organizes the genetic code's basic polarized pair of pairs (four bases of G, C, A, and T), and how it also organizes the universe's primal polarized pair of pairs (space, time, matter, and energy).

The *double p-tree (dp-tree)* can mirror its three levels of branches above with three levels of roots below. It has two domains, with 8 trigrams in each domain. Since the dp-tree holds a related family of polarities, the 8 chaos patterns above and 8 chaos patterns below can pair-bond across domains to develop 64 co-chaos patterns. They lock binary counting, addition, period-doubling, and exponential progression into a tight, fail-safe analinear synthesis.

THE DOUBLE P-TREE GROWS 8 TRIGRAMS ABOVE & 8 TRIGRAMS BELOW. THEY PAIR-BOND AS 8 X 8 PAIRS OF TRIPLETS TO FORM THE 64 HEXAGRAMS.

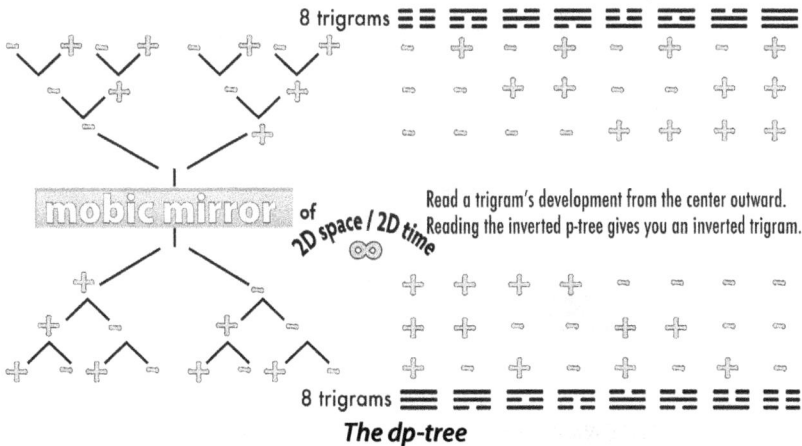

Read a trigram's development from the center outward. Reading the inverted p-tree gives you an inverted trigram.

The dp-tree

This paradigm establishes 64 hexagrams as polarized 6-packs that correlate mathematically with the 64 molecular 6-packs in DNA's double helix. They can unpack into RNA codons that automatically sort into their correct amino acid families, with I Ching messages reflecting their amino acid tasks! This paradigm also sets trinities and octaves echoing in all the physical sciences. It abracadabras octaves into chemistry's periodic table of elements, into subatomic baryons and mesons in physics, and even into octaves of music, the most mathematical of arts.

128 DNA Codons = 128 Trigrams = 64 Hexagrams

ANCIENT CHINA CALLED THIS THE XIAN-TIAN—
EARLY HEAVEN—PRIMAL ORDER OF TRIGRAMS.

128 DNA codons = 128 trigrams = 64 hexagrams

Now we have five distinct ways to view trigrams: (1) in unitized mode as binary counting; (2) in unitized mode as addition by 2s; (3) in analog mode as period-doubling and/or exponential growth; (4) as linear units and analog flow working together with a built-in nonlinearity that becomes analinear; (5) as 8 related, vertical period 3 windows of chaos dynamics developed on the p-tree at the third level of growth outward, above and below, which then can pair-bond into 64 co-chaos dynamics.

How can the progression from 2 to 8 carry so much freight? How can it generate our universe full of science's triplets, octaves, and other broken symmetries? It happens because the dp-tree's 8 × 8 = 64 co-chaos patterns ratify the universal paradigm. And the I Ching math can shorthand it!

5. Beyond the sturdy 4, we can't go home again

Between the 2 and 4 lies the hot-spot clout of the 3, succinctly expressed by a period 3 window in both its horizontal and vertical formats (hp3 and vp3). This series identifies the vp3. Yorke and Li identified the hp3 in *Period Three Implies Chaos*. Polymath Jack White called such 3-ness essential to his versatility.

The 3 automatically hold dynamic force, for good or ill. A trio of *anything* can compare and contrast its shifting bonds of relationship, much like a comic-strip triangle of humans who find themselves stranded on an island. Alliances form and reform, aligning two against one in shifting permutations...or all against each other...or even all 3 working together in common purpose.

Perhaps the ultimate example of the 3's archetypal power to work for the common good is the fact that it leads onward to the 4 archetype. The foursquare stability of 4 exists because the 2's path to 4 takes three different routes that remarkably and exactly echo the same numbers: *adding* $(2 + 2 = 4)$; *doubling* (2 doubled = 4); and *squaring* $(2^2 = 4)$. All three routes smoothly, easily, successfully merge their processes to reach the same goal of 4 without a hiccup. Thus the dynamic force of 3-ness led to the unparalleled stability of 4-ness.

In life, we often employ the stabilizing power of the 4, and it manifests most often as a *polarized* pair of pairs—for instance, as four compass directions, four human limbs, four partners playing bridge, four legs on a table, four wheels on a car.

Beyond the 4, however, such totally reinforced synchrony is forever lost. Progressing past the 4 opens a quandary over which route to take onward to 8. Addition's route of 2-4-6-8 must make an extra stop at 6, with its hiccup of slight delay along the way. But if 2's progression instead chooses analog routing to 8, that opens a new quandary: doubling or squaring? Or both?

Oh no, such woe! This plethora of choices turns relational analogs fuzzy past the 4, promoting vibratory resonances, echoes, and repercussions. In the curt, clear-cut domain of binary logic, an answer is either 0 or 1, *no* or *yes*. But analogs do not whisk along a computational 0-1 binary chute of forced choices. The reality we humans experience is so rich, multi-hued, nonlinear, even analinear, that it cannot be tidily cut, dried, and bundled into a standard, uniform, quantized, predictable harvest of life. Life's underlying skeleton of analinear math throws leeway into the fleshing out of events at every instant.

Yet that very leeway in the co-chaos paradigm as it progresses on beyond the 4 to the 8 is also what provides us with a measure of free will. Our choices can become wider and hopefully wiser than straightforward binary logic manages to dole out. The co-chaos patterning in your own continually flowing psyche, for instance, means that your relationships do not operate in clear-cut, binary choices. Instead, they hold tinges of ambivalence that can foster degrees of *yes* and *no* at the same time. Due to DNA and the mathematical paradigm that underpins it and indeed the whole universe, your life itself is analinear.

Beyond the 4's sacred, secure *temenos*, numeric progressions can never go home again to total resonant accord. This simple fact has a deep, unvoiced, and often unconscious profundity. We know it in our innermost core below words. No wonder Eden had four exits guarded by angels with flaming swords. No wonder Taoists stationed four gates beyond the sacred center of their rituals. No wonder Hindu yantras opened four doors to the outer world. No wonder the four horsemen of the Apocalypse rode to signal the end of the world... as we know it. On past the 4, number's methods and goals can never again mesh so sweetly into coterminous Eden. Oh, Paradise lost!

But if the 2's progression on beyond 4 can no longer enjoy the simple strength of a polarized pair of pairs, perhaps numbers can instead rise to a more sophisticated power? Indeed so, and this feat is showcased in the life-giving dynamic of the polarized pair of triplets. On the dp-tree, the 64 co-chaos patterns that generate all life, including that of the universe itself, come from the polarized pairs of triplets that pair-bond across two domains to form 8 × 8 co-chaos patterns.

At every level of our universe, the co-chaos paradigm provides life with its paradoxical stability combined with ever-emergent renewal. Through it, the universe thinks, feels, aspires. This paradigm is profoundly elegant, yet so simple that the Chinese notated it in I Ching shorthand circa 5,000 years ago. As Paul Davies remarked in *The Cosmic Blueprint*: "The fact that the universe is full of complexities does not mean that the underlying laws are also complex."

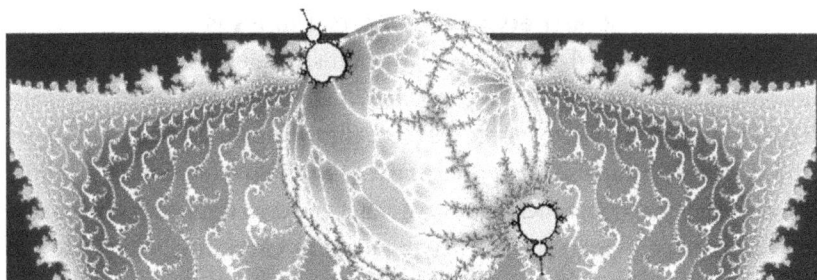

Chapter 14: Numbers With Heart

1. Fractal nesting

The great hologram of our ever-emergent universe is not a mere chance pile of events to be dismissed as meaningless. We each move uniquely, often unconsciously, in huge and tiny self-similar patterns that are evolving even as they also keep repeating their continual slight variations. Meanwhile, our conscious, linear science marches on its intentional study of cause-and-effect.

This universal hologram is woven out of numbers, but it's an analinear, oddly relational kind of numbers that create nature's fractal patterns. It connects the dots of our days into synchronicities hidden within the spacing and timing of everyday life. But when our human minds notice the synchronizing work of universal mind in nature, logic is often unwilling to notice it.

Yet there are clues if you choose to be aware of them. For instance, Jungian psychologists know that mandalas trigger a symbolizing, organizing, healing impetus in the psyche. If a client starts to work with mandalas by drawing them or by walking in a mandalic maze, psychological reorganization and healing are likely to occur. That's why we unconsciously seek out a rose window, pause in the still center of a maze, or stand to admire the Arc de Triomphe's hub-like view.

How reassuring it is to realize that mandalic patterns such as the endless Mandelbrot Set are embedded in nature itself! Just when Earth's ecosystem is becoming sorely blighted by a techno-mad, money-focused, world-wide culture that's losing connection with this planet's natural environment, infinite mandalas meanwhile are being found in math's chaos dynamics. Can it be that nature is trying to cure itself by revealing fractal manifestations of its vast hidden beauty to those who have damaged it most—us humans? Is it possible that nature seeks to heal itself from within by revealing this fractal geometry to our emerging global society that is losing touch with the natural world?

Shades of the folk wisdom that honored Mother Nature! Shades of Taoism with its nesting boxes of events that hover beyond ordinary logic! Shades of synchronicity, Carl Jung's acausal connecting principle where random-seeming events mesh into patterns of hidden meaning.

2. Archetypes: Apollo vs. Dionysus

Patterns are buried in the welter of human thoughts. In the personal unconscious, they reveal clusters of psychic energy that Freud called *complexes*. Jung emphasized a collective aspect of those clusters under the name *archetypes*. Archetypes are complexes organized in the collective unconscious. They hold sway whether you notice them or not, whether you even think they exist or not.

For instance, Apollo is an archetype noted in Greek mythology. He was typically depicted as a handsome, beardless youth with light hair and eyes. Apollo drove the chariot of the sun across the sky to shed clarifying light on the world. He played the calming lyre, was loyal, responsible, and appreciated his mother.

You know this kind of person…the calm, clear-headed, and knowledgeable person who likes to keep life controlled, defined, and tidy, who does not want to stray beneath the surface to examine life's dark, roiled underpinnings. The Apollo archetype is much honored in Western culture. My formal education trained me to honor and emulate this logical, responsible mindset. Got a problem? Apollo is the logical strategist. An Apollonian approach can solve it, fix it, or end it.

But there's more to life than Apollo's archetype. For some, moving beyond cool, calm logic means running away from Apollo straight to Dionysus, another Greek god (in Roman mythology, the more dissolute Bacchus). Dionysius was depicted as bearded or long-haired, or both, with dark hair and dark eyes. His archetype evokes passionate release into the emotional wilds beyond logic.

Dionysius likes music, too, but not the lyre. A very different kind. He is the god of hypnotic music, flowing wine, ecstatic dance. Of exaltation, terror, and altered states. He liberates his followers from self-conscious cares and fears; he breaks the shackles of oppressive restraint. He loves revelry, drink, and drama… especially melodrama. His enthralled followers succumb to trance-like rhythms that lure the writhing mob into tempestuous, thrilling release.

Triumph of Bacchus and Ariadne by Carracci Annibale 1560 – 1609

The Dionysian archetype is an emotional powerhouse. It can enslave or liberate. In extreme passion, people lose self-control. They drug, rape, fall deeply in love, live for music. They may become rave devotees freed up by dance, sex, drugs. Some use wild weekend parties to counteract a humdrum office cubicle. Moments or years later, a person may say bemusedly, "I don't know what came over me," or "Something possessed me," or "I felt driven!" or "The devil made me do it," or "I followed my bliss, and it blessed me."

Apollonian logic shrinks away from Dionysian release. It dreads the psychic act of giving over to passion; it fears the loss of self-control. It avoids relinquishing personal power to what it sees as brutal gods and canny, greedy priests, as wizards, healers, and gurus who manipulate their followers, if given the chance.

Thus the Apollonian fear of passion tamps life down to keep it orderly, but the Dionysian fear is a lack of passion! It ramps life up to honor passions. Apollo and Dionysius are just two of many archetypal patterns in the collective unconscious. From among them, your own unique identity will pick out its major motifs and then assert them again and again. You call it your personality.

Your personality asserts itself in ways large and small, never with the exact same details, and always with a potential to evolve along the timeline of your existence. It is a living system in continual flux, full of habits, feelings, wishes, memories, dreams, reflections. The dynamics of co-chaos shape it as surely as they shape your material lungs, feet, and face…but your invisible psyche's complexes are evident only in your words, motions, and emotions…so your complexes are harder to chart than your face.

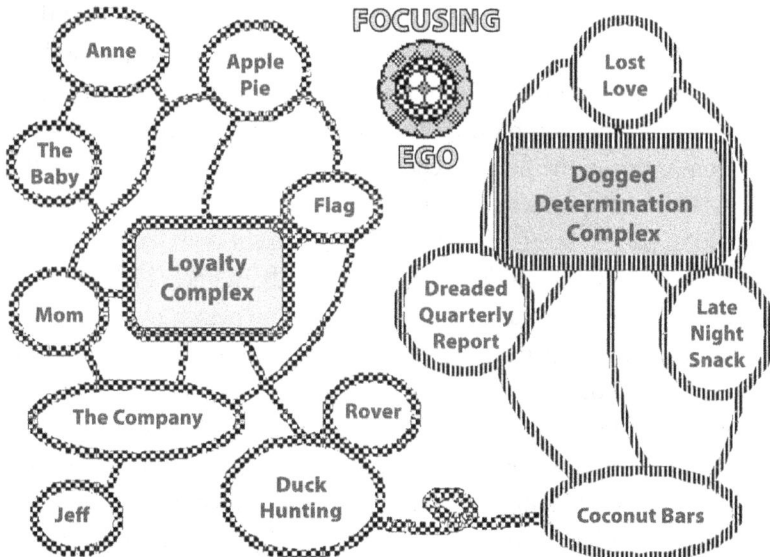

Two of Joe's complexes in relationship

Your psyche is a living system in continual flux. The psyche's behavior comes from complexes of energy that are devilishly hard to chart, simply because they are visible only in words, motions, and emotions.

Psychological complexes can be helpful to you, or the reverse. Or both. For instance, a good habit of persevering in the face of difficulty can take you far...until you adamantly refuse to leave a sinking ship. Biting your nails can assuage your munchy habit even as your vanity condemns it.

So how do you tweak the counterproductive aspect of a complex into something more helpful? Evolve it! As your complex generates its typical, identifiable behavior that locks your life into repeating its dynamic in ever-new variations on scales big and small, always with unique specific details along your timeline, try to observe it, adjust it to lessen any counterproductive aspects. Seek to phase out the negative while keeping what is useful. Is that really possible? Yes!

Consider this: the co-chaos patterns in your psyche shape your reality as surely as co-chaos patterns in DNA shaped your lungs, feet, and face. However, your psyche's invisible complexes become evident only in your words, motions, and emotions. That makes them harder to chart and predict, and even more so since your psyche's behavior is analinear, with a built-in measure of free will.

So instead, let's talk about a visible version of nonlinear behavior that is easily tracked to observe its changes. The Mandelbrot set has transitional moments in its borderline conditions, phases where a current pattern modifies. The psyche has them, too. They readjust where you stand in a pattern. Maybe you're in the midst of losing your smoking addiction/great job/excess weight/best friend/ fear of flying/good health. There's no secure boundary here. You miss the old footing. You whirl in impulses that pull you momentarily this way and that.

Right now, your psychological condition is not just *either/or*, but instead, it's something more complex. It's in a transitional phase because boundary conditions occur in the fluid psyche as well as in the Mandelbrot set. Being pulled about by the attractors of a shifting psychological dynamic is akin to a Mandelbrot pattern being tweaked by strange attractors and strange repellers.

You can cultivate a mindset that champions both linear logic and analog relationships, that strives for an overview lifting your perspective up enough to see more of your life's pattern. Then you can spot a good direction to take. It moves you on beyond those furiously fragmented boundary conditions into a more assured, calm region of life with less melodrama and more true comfort.

But even then, each day still iterates a new variation along your lifeline. It never retraces the old path precisely. The chunk of each day feeds back into the mix and propels you onward...much as the Silly Centimeter vine grows in Volume 1. Each day's result feeds back into your life and moves you onward.

3. Analog emotion runs ahead of linear logic

The equation of your life becomes an ongoing process, not a final solution. The constant in it is you. You inhabit a moving design of possibility that you cannot even see very well from your position riding on the arrow of time in this 3D space bubble. But occasionally, the bandwidth of your consciousness may dilate enough to receive from the other bubble an amazing overview of your life's patterns broadening in 3D time.

If you do, your body is likely to respond forcefully, even quicker than your thought. Brain scientists know that a reaction happens in your body before your conscious mind even registers what has triggered it in the brain's cortex. In other words, your body reacts before a thought notifies you about what you're already reacting to. You are unconsciously reacting before you even know why!

The frustrating thing with analog resonances?…they're so hard to evaluate in a merely logical fashion. Something just rings your chimes, turns you on, tunes you in. You feel its vibes. A subliminal cue rises and resounds throughout you, around you. It sweeps you along in a moment's majestic, poignant, witty, fearful, glorious, awful, cosmic swell…that informs you somehow.

Moreover, all your senses—sight, sound, touch, taste, smell—trigger subjective feelings and objective thoughts. As your senses report their data, your emotions are delivering their own subjective, qualitative commentary: "That noise is too loud!" "Oh, I like the green one!" "It's really good chocolate!"

Joseph LeDoux of the Center for Neural Science at New York University studied the *amygdalae,* two almond-shaped clusters in the deepest part of the limbic system at the brain stem, where dreams and other unconscious data are processed to help your brain's system of unconscious motivators do their work.

He realized each little almond-shaped *amygdala* is part of your unconscious threat detection system. It also operates in your systems that process the significance of stimuli related to eating, drinking, sex, and addictive drugs.

(Fractal trivia: *Amygdala* was the insider joke name on a newsletter about fractals, published by Rollo Silver from 1989 to 1995. Silver chose that name *Amygdala* because…drum roll here…the amygdala is shaped like an almond, and he chose it to honor Benoit Mandelbrot, whose last name in Yiddish means *almond bread.*)

How do you study the analinear mind, not just a material brain? The holistic living system of a psyche demands something beyond a biology lab's techniques of dissecting, sorting, weighing,…and even beyond psychology's experimental techniques that chop human nature into linear chunks, offering packets of statistics distributed across columns to dispense blips of information.

Okay, yes, such methods work quite well in aspects of many sciences. But what about the holistic system of you, a living person whose psyche renders you more than the sum of your parts? The code that made your body and mind is analinear. The leeway in its numbers is what allows you a certain measure of free will…and you do somehow sense that you've got some choices, right?… even if Laplacian causal objectivity insists that determinism is quite logical.

4. Succor? Sucker?

Speaking of logical versus illogical, many people today think the very idea of divinity is dead, since divinity isn't quantifiable, accountable, or visible. Yet those same people may still exclaim "Goddammit!" or just "Dammit!" Is it just cultural habit, or do they subliminally realize they've referenced a generating, organizing dynamic that permeates the universe as ubiquitously as atoms and isotopes, as implicitly as space and time? It seems the idea of divine force did not really die out, but for many, the image of a God who cares has just collapsed away.

Good riddance to God, some say. A primitive superstition. Many tacitly agree with Karl Marx's view that religion is the opiate of the people. But interestingly, as religions have faded in society, stimulant and narcotic use have gone up, adjusting (temporarily) the human spirit. Many no longer have the notion of God to nurture the spirit by trying to act better, love better, live better, so they instead seek and honor the petty escapist gods of addiction.

Today the escape into addiction is often triggered by a search for psychic release from the data-mad rush of daily events. People climb an alcoholic or crack or opioid-induced stairway into a glassy mezzanine of heaven. They become addicts of drugs, food, sex, virtual reality, danger, even work…anything to minimize the ache of a life without the lofty spirit that we moderns have finally managed to engineer ourselves past believing in.

Addictions offer a pseudo-release that is a poor imitation of what spiritual faith once brought. They are a spurious substitute for the elevation of spirit that comes from contrition or fasting or meditation or ecstatic dance or charity work. An escape into addiction promises a momentary release for the beleaguered animal spirits, but it lacks the uplift into transcendent spirit. Addiction offers an escape *from*, not a road *to*…which finally becomes just another dead end, not a path to enlightenment, unlike the spiritual bridge to the divine that is the goal of mystical seeking.

At some deep level, we know that the war on drugs is more rhetoric than result, that we have met the enemy, and it marauds within. It derides our hunger estranged from higher meaning in this modern, soulless, data-driven world. Bombarded by too much information, strafed by too many problems, juggling too many worries, we hunger for some kind of transcendent, spiritually relevant meaning. We sicken without it, without knowing why.

As we're developing marvels of technology to "Plug it, play it, burn it, rip it, rap it, sneak it, drag it, drop it, zip it, zap it"—aspiring toward heights of tech nirvana, we've also lost some rough Eden of connective wholeness. Bereft of our kinetic, unspoken relationship to nature, to higher meaning, now in some half-conscious way, we fret over what we've lost and do not even know how to name it. We surfeit our flesh with trinkets and starve the ineffable soul.

My brother-in-law Jack once joked that the word *ineffable* is just Latin for *unfuckable*. Okay, and *psyche* is just Greek for *soul*. Our current culture suffers from atrophy of the soul. Ill with anorexia, it hungers for divine meaning. Carl Jung diagnosed that malady in 1933 in *Modern Man in Search of a Soul*.

Where can we turn for our soul's nourishment? Emaciated, where can the soul succor that spiritual yearning? Succor, indeed! Some part of us today yells, "Sucker!" Soul-starved, must we gulp pills, shoot up? Are we so desperate to regain connection with spirit that we must swallow religion's historical intolerance of other religions? Where do we find an answer that will not betray us, traduce others, straight-jacket the divine possibility into a limited, culture-bound vision?

5. The I Ching is not a religion; it is a mathematical/spiritual tool

To quote *Merriam-Webster* and Wikipedia, "Religion is a social-cultural system of designated behaviors and practices, morals, world views, texts, sanctified places, prophecies, ethics, or organizations, that relates humanity to supernatural, transcendental, or spiritual elements. However, there is no scholarly consensus over what precisely constitutes a religion." Whatever their origin, all religions hold tenets and principles that are codified for the many by the few. Followers are expected to believe the teachings, even if they have doubts.

Uh-oh! I always have doubts about something that will not let me doubt it, explore it, test it, evaluate it. I've doubted my own God dreams. Who wouldn't? That's why I made such an effort to find out if DNA and RNA really do correlate with the I Ching's math and meaning, why I've tried very hard to see if our science can actually support the concept of a Double Bubble universe where the hidden, mirror-twin bubble holds the lost pole of gravitation.

Perhaps religions are currently losing favor because they are not scientifically testable in labs. Myself, I especially appreciated how I could test the I Ching's answers against reality over time. For the first 6 months or so, I often doubted an I Ching answer to my question on how the upcoming day would be because I did not see how it could predict the day's events or attitudes.

I had to wait and see what played out over time. I kept a journal of my questions, answers, and thoughts, so despite my slow learning curve on how to interpret hexagram answers funneled through the mindset of an archaic, agrarian, feudal Chinese culture, eventually I admitted that the I Ching answers always turned out to be relevant, and in fact, they sometimes were more pointedly correct than my modern, ego-defended mind wanted to concede.

6. My answer is the I Ching

With the I Ching, I can sit anywhere at any time and divine. Literally. Not everything, no, but the co-chaos patterns of my life, yes. The I Ching is the most solitary and meditative of oracles. It is always one-on-one…I sit with it *tete-a-tete*…so that something in it is finally impervious to the approval or disapproval of society. It's just me sitting there, trying to divine the unvarnished truth about something that is on my mind. The I Ching always shows me new angles of contemplation on my issue.

Of course, I can mess it up by botching my response. The subjective stance I take is already a given, coming from my own limited niche in the vast reality. Defensiveness, Derisiveness. Denial. But time straightens me out. So I stand in an answer and look around to get a new perspective on some issue. With it, I transcend my limits. I reach past my frailty to glance at a core of truth beyond the projective screen of my ego, experiences, complexes, whims, desires, and mere chance. My records over more than 35 years show me that the I Ching helps me understand things better, helps me make better choices, deconstructs my troubles, and realigns my impulses toward a better way.

How can that be? Life has its reasons that linear logic does not know. Analinear ones. Clear-minded Apollo vs. passionate Dionysius—they typify the split-brain perspective that has bedeviled Western thought for more than 2,500 years. It split us into classicist vs. romantic, conservative vs. liberal, right vs. left, hard head vs. bleeding heart. And shadow hides within the divide.

It is possible to encompass polarities in a larger, transcendent third way that embraces both poles and reconciles them. It is epitomized by the I Ching that can shorthand co-chaos patterns in our DNA and in the master code. By combining the linear shunts of binary chunks and the analog webs of cycling relationships, this analinear patterning takes all life somewhere new. Its peculiar beauty weaves together mind and matter, math and morality…and all of this intrigues and enchants me. It heals the splits in me and directs my issues to a wisdom that resides in nature itself.

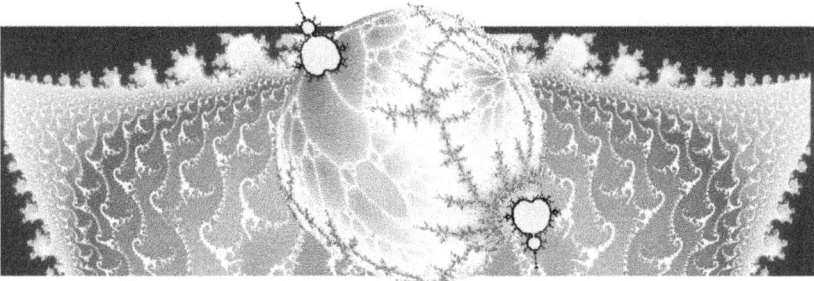

Chapter 15: Wheels of Trigrams

1. The binary order of trigrams

In Chapter 13, you saw how the trigrams on a p-tree count from 0 to 7 in binary sequence. Shao Yong 9 (1011–1077 CE) found this binary order of trigrams. Ancient scholars called the order *xiantian*, meaning it was original, ideal, pristine, uncorrupted, *a priori*. This order's name of *xiantian* is translated into English variously as the order of Early Heaven, Before Heaven, Primal Heaven, World of Thought, Old Family, and other names…but for our purposes in this series, we'll call it the binary order.

Scholars considered the *xiantian* row of trigrams perfect because it counts in binary sequence. If the 8 trigrams are arranged on a *xiantian* (binary) wheel, they sit complementary, in mirror-opposite pairs around the wheel, with each pair showing a polarized balance of 3 yins and 3 yangs.

Here the trigrams sit binary order in a row and also on a trigram wheel.

0	1	2	3	4	5	6	7
Black Earth	Purple Mountain	Blue Water	Green Wood-Wind	Yellow Thunder	Orange Fire	Red Lake	*White Heaven*
Mother	3rd Daughter	2nd Daughter	1st Daughter	1st Son	2nd Son	3rd Son	Father

Trigram wheel in binary, xiantian order (the ideal, primal order)

Since the binary order of trigrams and hexagrams was considered perfect, it often decorated homes, community buildings, and especially tombs. The binary order was also widely displayed in everyday society and used in mathematics, philosophy, medicine, martial arts, and astrology.

Its originator, Shao Yong, was a Song dynasty monk, philosopher, poet, and historian. Although neo-Confucianists embraced him, Shao Yong is often considered Taoist because of his humility, his refusal of government positions, and his "image-number study" approach to using the I Ching. His biography is recorded in the *Sung Shi* (*History of the Sung Dynasty*).

2. The analog order of trigrams

When Shao Yong developed the binary order of trigrams and hexagrams, an older, well-known order had been in place for two millennia. It was the analog, relational, *houtian* order. This order's name of *houtian* is translated into English variously as the order of Later Heaven, After Heaven, World of Action, King Wen, Chou I, New Family, etc. But for our purpose here, we'll call it the analog order.

The same trigrams, of course, appear in either order, analog or binary, but they sit in different positions on the row or wheel. It's as if the two families of trigrams live near each other, they are neighborly, yet they set up their households differently. Each family has two parents and six children…three sons and three daughters. Those are the similarities. What are the differences?

Here the trigrams sit analog order in a row and also on a trigram wheel.

Black Earth	Green Wood-Wind	Orange Fire	Red Lake	Purple Mountain	Blue Water	Yellow Thunder	White Heaven
Mother	1st Daughter	2nd Daughter	3rd Daughter	3rd Son	2nd Son	1st Son	Father

Trigram wheel in analog, houtian order (the real, manifested order)

The analog order of trigrams and hexagrams was first written down by King Wen and his son, the Duke of Zhou. They were co-founders of the Zhou dynasty (1046–256 BCE), the longest-lasting dynasty in Chinese history. English translators also spell the Zhou name as Chou or Chao. It's droll that the Chao dynasty's two co-founders should be the first to record co-chaos patterns.

In analog order, the trigrams on a row no longer count in base-2 binary numbers from 000 to 111, or in base-10 digits, from 0 through 7. Instead, Mother *Earth* still sits protectively at the left end of the row, and Father *Heaven* still sits at the right on the other end. But all six other trigrams have changed places in the row. Moreover, four of the siblings—*Mountain Water, Fire,* and *Lake*—have even changed their birth order and switched gender!

This analog order is cycling, searching, and innovative. It is used in *feng shui* because this order is thought to show the life force that tracks living rhythms and stirs latent energies, stimulating them into movement and growth.

3. Comparing binary and analog orders on two color wheels

A more scientific way to compare the two trigram orders is to let yin-yang math determine their trigram positions on two different color wheels, the paint wheel and the light wheel. We've developed two different kinds of color wheels because our human eyes use light to perceive things in two different yet complementary ways. (Weirdly, we also have two complementary theories of color vision: trichromatic theory and opponent-process theory. Both theories work for how our brains process an image; they just work differently.)

1. PAINT'S REFLECTING COLOR WHEEL OF PRIMARY COLORS: *BLUE, RED, YELLOW*...

We can only see most objects if light rays bounce off of them. Any light-absorbing object...like your T-shirt, a dog's fur, an apple...will *reflect* any unabsorbed light because it bounces off the object into your eyes.

When you are outdoors on a sunny day, rays of white light from the sun will shine on the apple in your hand. The apple looks red because its skin reflects only the red light wavelengths and absorbs all the other waves from the white light hitting it. Any leftover, unsubtracted rays (in this case, red wavelengths) are not absorbed and reflect into your eyes. You see red.

2. LIGHT'S PROJECTING COLOR WHEEL OF PRIMARY COLORS: *BLUE, RED, AND GREEN*...

However, we see some objects because they generate and project their own light...for instance, the sun, a TV screen, a car's headlights, fireflies, or the neon sign hanging outside a bar. They *project* their colored light right into your eyes.

Now let's correlate the two different color wheels with the two different trigram orders. Then we'll compare the results. Trigrams can help us see how the two different color wheels correlate mathematically.

Below on an artist's paint wheel…the trigrams are arranged in binary order, with both parents in the center. All six children gather around them, sequenced simultaneously by color, birth order, gender, and binary counting.

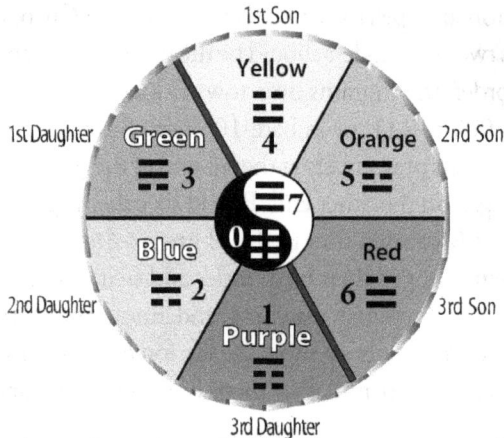

Paint's reflecting, subtractive color wheel puts trigrams in binary order

Note that due to the symmetries in the binary order of trigrams…

First: Trigrams in binary order correlate to paint's reflecting color wheel of the artist. Its primary colors of blue, yellow, and red sit in a triangle on the wheel.

Second: The trigrams on the paint wheel sit across from each other as counter-balancing complementary colors. For instance, if you stare at purple ☶, you get an afterimage that is yellow ☳ across the wheel. Ditto for red ☲ across from green ☴. Also for orange ☵ across from blue ☰ . Each cross-pair on the reflecting color wheel become each other's afterimage.

Third: Each cross-pair also provides reversing mirror-images of nature, for they express each other's poetic afterimage. Purple *Mountain* rises up, but yellow lightning brings *Thunder* bolts. The flame of orange *Fire* burns, but the moisture of blue *Water* extinguishes. At sunset, the red *Lake* has fish that swim, but the green *Wood-Wind* has birds that fly at dawn. *Heaven's* transparent air lets sunlight reveal whatever it touches, but *Earth's* black opacity cloaks whatever it holds in pregnant mystery.

Fourth: All four trigrams with symmetric lines are cross-paired to balance each other's polarity. Thus the cross-pairs of *Heaven* ☰ and *Earth* ☷ are polar opposites. Likewise for *Fire* ☲ and *Water* ☵. Moreover, these four trigrams with symmetric lines may have even been flipped upside down, too, but we cannot be certain since they'd look the same, flipped or not.

Fifth: The four non-symmetric trigrams are cross-paired as flip-flopped images of each other's polarity. Thus *Lake* ☱ and *Wood-Wind* ☴ turn each other's polarity upside down. Ditto for *Mountain* ☶ and *Thunder* ☳.

Below on a physicist's light wheel...the trigrams are arranged in analog order, with both parents in the center. All six children gather around them, sequenced by color, a new birth order, maybe a switched gender, and no binary counting.

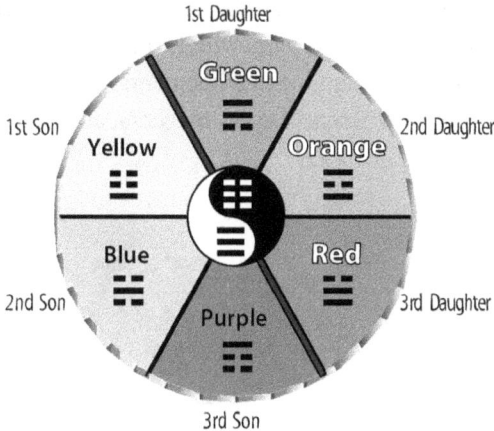

Light's projecting, additive color wheel puts trigrams in analog order

Note that due to the symmetries in the analog order of trigrams...

First: Trigrams in analog order correlate to light's projecting color wheel of the physicist. Its primary colors of blue, green, and red sit in a triangle on the wheel. The two parents in the center have exchanged positions on the wheel, as have the first son and daughter. The other siblings have also switched gender.

Second: The four trigrams with symmetric lines are still cross-paired to balance each other's polarity. We still cannot tell if they've also flipped upside down, for they'd look the same either way. Thus clear *Heaven* ☰ and opaque *Earth* ☷ still reverse each other's polarity. Ditto for orange *Fire* ☲ and blue *Water* ☵ (symbolizing active water, often called *Canyon* or *Abyss*).

Third: On the light wheel, the four non-symmetric trigrams are still cross-paired as flip-flopped images of each other's polarity. If you examine red *Lake* ☱ and green *Wood-Wind* ☴, for instance, you find that they reverse each other's polarity. This likewise holds true for purple *Mountain* ☶ and the yellow crack of lightning that brings *Thunder* ☳.

4. Comparing the two color wheels

What is the purpose of comparing the I Ching's two different trigram orders with our two different color wheels, paint and light? My goal is to show you that a shared master code underlies both systems. Both systems—trigrams and color wheels—are following the same elaborate, synchronized, mathematical dance. They employ an underlying master code paradigm that is inherent in the universe itself.

Now let's set the two color wheels side by side for an overview. Each wheel has three primary colors. On paint's reflecting, subtractive wheel, the primary colors merge to black. On light's projecting, additive wheel, the primary colors merge to white. Below each wheel is a cube. Its three visible sides show the three primary colors of that wheel. (The paint wheel's three primary colors were once called blue, red, and yellow, but due to refinements in color printing, they got updated to the more exact terms of cyan, magenta, and yellow.)

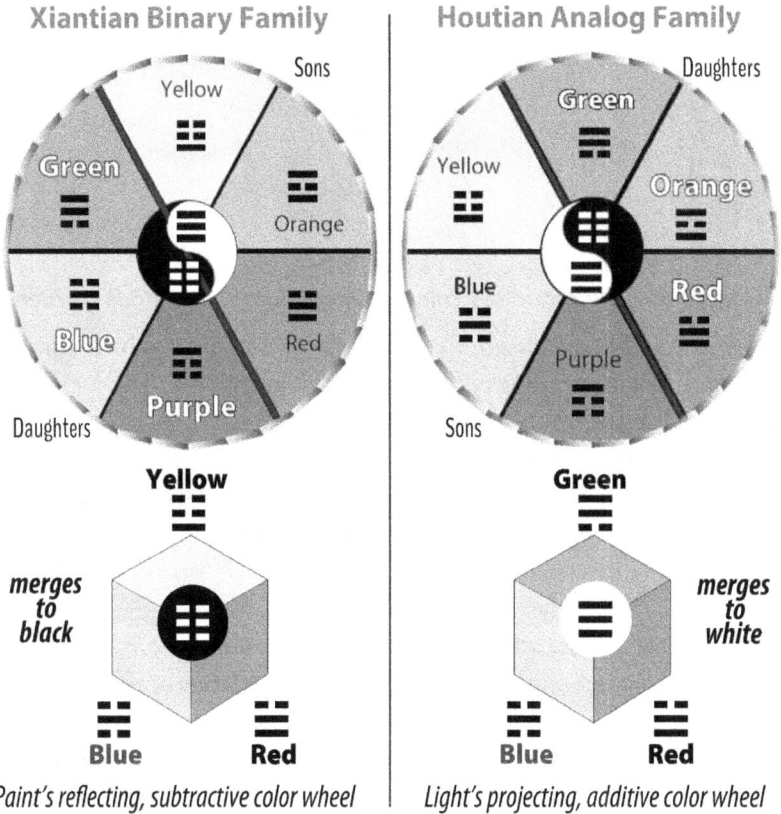

Comparing the two color wheels of paint and light

Other relationships between trigrams and nature will appear as you explore their mathematical fit. Hellmut Wilhelm said in *Heaven, Earth, and Man in the Book of Changes*, "The system of *The Book of Changes* is the representation of a multidimensional world. Pairs of opposites should not be looked for only at the poles of a one-dimensional axis. Depending on the direction of view, there will be found a number of different opposites for every given concept or situation."

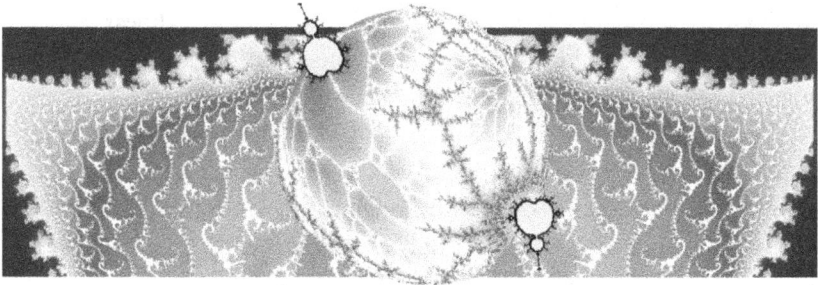

Chapter 16: Remixing History

1. The "chicken & egg *puzzle*" of xiantian & houtian

To convey a bit of the I Ching's historical depth, let's consider the question of which order of trigrams and hexagrams came first. Was it the binary order of *xiantian* (Early Heaven)? Or was it the analog order of *houtian* (Later Heaven)?

You might assume that Early Heaven's binary order must have appeared in Chinese history before the appearance of Later Heaven's analog order. But, no, such was not the case.

In the 1970s, a modern study of the I Ching revealed that Chinese scholarship had long suffered under the false impression of which came first, the "chicken" of binary *xiantian* or the "egg" of analog *houtian*. It gave a shocking rebuttal to a previous millennium of I Ching scholarship.

Archeologists by now have definitively proved that King Wen's houtian order of Later Heaven appeared in historical documents several millennia before Shao Yong's xiantian order of Early Heaven showed up.

So why those misleading, time-skewed names for the two orders of trigrams and hexagrams? In the 12th century CE, Chinese scholars spread much historical confusion by naming the binary order *xiantian* (Early Heaven) and the analog order *houtian* (Later Heaven). Why did they do that?

When Shao Yong presented his new binary arrangement of trigrams and hexagrams in the 12th century CE, Chinese scholars viewed it with delighted awe. They considered it so beautiful, so pure, so mathematically elegant that they decided Shao Yong's binary order was worthy of having been composed by legendary Emperor Fuxi (circa 3000 BCE), who was said to have first discovered the trigrams. So they gave Shao Yong's binary order a title that implied it had originated first.

Francois Louis explains it this way in *The Genesis of an Icon: The Taiji's Diagram's Early History*: "By the twelfth century the *xiantian* trigram order was hailed as a sublime, naturally perfect arrangement, for it provided the most systematic known way of organizing the trigrams. To leading twelfth-century intellectuals, most notably Zhu Xi, this order, whether in its linear sequence or in the inverted pairing of the circle, represented a natural principle of organization, corresponding to the original conception of Fu Xi."

Those awed scholars decided to rearrange I Ching history by rebranding Shao Yong's new binary order. They called it the *xiantian* or Early Heaven order, thus identifying the binary order as the "original, inborn, natural, innate" order.

The scholars made the remix complete by also rebranding King Wen's long-familiar analog order, which had been used to consult the I Ching oracle for over 2,000 years. They bestowed upon it a brand-new title: *houtian*, meaning Later Heaven order. By establishing those official titles for the two orders, the old was made new, the new was made old, and scholars managed to reshuffle 2,000 years of I Ching history.

Xiantian and *houtian* are both compound names made of two characters each. They share a word in common—*tian*. By itself, this word *tian* translates literally as *sky* or *heaven*, with an inference of coming from higher law or order. The *Encyclopedia Britannica* reports, "Scholars generally agreed that *tian* [heaven] was the source of moral law, but for centuries they debated whether *tian* responded to human pleas and rewarded and punished human actions, or whether events merely followed the order and principles established by *tian* [heaven]....Chinese rulers were traditionally referred to as Son of Heaven (*tianzi*), and their authority was believed to emanate from *tian*."

Two other words exist in those compound names: *xian* and *hou*. They may be translated alone as *early* and *later*, but when compounded in the fashion of *xiantian* and *houtian*, their meanings morph into something more like "ideal, primal, original way of heaven" and "real, manifesting way of heaven." In short, its underlying concept is that there's many a slip twixt the ideal and the real.

When that official rebranding of the I Ching orders occurred in the Imperial court circa 1050 CE, soon other scholars heard and read the titles of *xiantian* and *houtian* so often and so officially that within perhaps 100 years, those names were assumed to be accurately representing the historical timeline of the two orders of trigrams and hexagrams. Thus, due to the misleading names of those two orders of trigrams and hexagrams, Chinese scholars for the next millennium were misled by a scholarly reshuffling of history.

It certainly fooled me until I had the good fortune to come upon some more recent archeological findings. It was a major reason why I updated the series in this new edition. I hope to correct any wrong assumptions people may have made from the names I'd used previously. I changed the English names I'd formerly used—Old Family and New Family orders—into binary and analog orders. I did not want to get caught again by those labels that refer to time.

And who knows? Time-wise, our "modernist" history may go topsy-turvy again due to a future archeological dig. For instance, who knows what wonders may come from the massive tomb of Qin Shi Huang near Xian?

I also conjecture that reshuffling and rebranding both hexagram orders a thousand years ago must have produced some cognitive dissonance during that transition. Francois Louis said in *The Genesis of an Icon: The Taiji's Diagram's Early History*, "…intellectuals like Lou and Zhu were well aware of the relative novelty of the *xiantian* symbols, yet insisted on their primordial perfection and timelessness, which resulted in the paradoxical tension in their texts between the simultaneous modernity and antiquity of the diagrams….

"…the manufactured antiquity of the *xiantian* trigrams was a means of promoting the new arrangement [binary order]….In fact, the adoption of the suggestive terms 'Before Heaven' and 'After Heaven' to distinguish the two trigram cycles was so effective that it eventually led to the anachronistic perception that the younger sequence was the older one; and vice versa."

2. Fractal reality–a shared understanding of the world?

The *xiantian* order is the binary sequence that Leibniz recognized in a woodcut sent to him in 1701 by Joachim Bouvet, a Jesuit missionary to China. Leibniz was greatly influenced by the I Ching's mathematics, harmonic balance, and philosophy, which he assumed to be Confucian.

However, the I Ching's philosophy is Taoist, not Confucian, and it was developed long before Confucius, who lived 500 years after King Wen first wrote down the I Ching while sitting in jail around 1060 BCE. Some scholars even dare to claim that Confucius probably never saw the I Ching.

Leibniz praised the I Ching so highly in print that French philosopher Voltaire responded with the novel *Candide* (1759). In it, he satirized Leibniz with witty brutality as Dr. Pangloss (the name means *all-tongue*). Pangloss is portrayed as a blithering idiot professor of "métaphysico-théologo-cosmolonigologie." Throughout the book, amidst dire predicaments and great sorrows, he keeps proclaiming with foolhardy optimism to the book's hero, Candide, "All things are for the best in this best of all possible worlds."

As biting satire, the book immediately became popular. It was mainly funny due to Voltaire's caricature of Leibniz as Dr. Pangloss, who sententiously kept spouting a distorted version of Taoist philosophy to the naive hero, Candide.

In fact, Voltaire was mocking not only Leibniz but also Jean-Jacques Rousseau, a French philosopher whose ideas on arts, sciences, and the "natural human" made him one of the most influential Enlightenment thinkers. Rousseau's regard for nature strongly influenced Western governmental, social, and aesthetic values.

To a witty and facile wordsmith like Voltaire, the unconscious, wordless way of the Tao might appear stupidly naive, yet by the end of Voltaire's fantastical book, its hero Candide has somehow managed to settle into a surprisingly mundane, natural, and realistic happiness—shades of Taoism!—by cultivating lemons and pistachios in his own garden.

At the book's end, Dr. Pangloss says to Candide, "There is a concatenation of events in this best of all possible worlds, for if you had not been kicked out of a magnificent castle for love of Miss Cunegonde…if you had not been put into the Inquisition…if you had not walked over America…if you had not stabbed the Baron…if you had not lost all your sheep from the fine country of El Dorado, you would not be here eating preserved citrons and pistachio nuts."

"All that is very well," answered Candide, "but let us cultivate our garden." Thus, remarkably, despite the book's many outrageous adventures and that ridiculous (and ridiculed) Panglossian bombast, by its end, Candide has found out how to reside contentedly, Taoistically, in his own little Garden of Eden.

As the Enlightenment was gaining momentum in Europe, Leibniz championed the I Ching. His act was broadly viewed as a bold challenge to the Catholic Church's authority. However, Leibniz was strongly committed to upholding religion and royal authority, and he even seemed hopeful that the Catholic and Protestant churches would reunite.

Leibniz did not see the I Ching as a religion. He said an irreducible core of math hid in its metaphysical depths, and its combination of math and meaning showed a "natural philosophy." Leibniz hoped it would someday become part of a new "universal language" offering a shared way of understanding the world.

I agree. We need a shared way of understanding the world. A global society's shared language will span math, science, technology, culture, customs, ethics, and philosophy. Hopefully, it will manage to balance the greater good with individual rights. Although economics and politics must play an integral part in the new global reality, so must honoring human worth and meaning in life.

The I Ching offers all that and more. Its grasp stretches from math to philosophy, from ancient past to far future, from science to serenity. Its shorthand goes deeper than words. According to this TOE, the 64 hexagrams can shorthand a master code that generated the ongoing life of this universe. It also templated the genetic code variant that made possible our own fleeting lives. The paradigm's fractal power to organize life abounds visibly in us and around us, from fossils in rocks to *begets* in the Bible.

At the atomic level, a variant of the paradigm puts octaves into the periodic table of chemical elements. At the quantum level, subatomic particles group in the Eightfold Way, and quarks in their triplets echo a variation on the I Ching's trigrams. "The Eightfold Way is to elementary particles what the Periodic Table is to chemical elements," said Physicist Sean Carroll in the *New York Times*.

Way on down at the far-tinier mobic scale, co-chaos rhythms generate space and time. Then they set matter and energy into that dimensionality to produce the coalescing patterns of all events. This series' next book, *Tao of Life*, describes how the I Ching's math shorthand and philosophical text correlate with the genetic code that originates all the bits of organic life in this bubble.

3. Hexagrams offer math & philosophy

A true Theory of Everything would include not only the four primals of space, time, matter, and energy in their physical, scientific, and quantitative values, but also the realm of mind, which is notoriously full of intangible, qualitative values that are far more difficult to name and evaluate. The I Ching's math and philosophy can span that gap, but utilizing its depth requires a calm grounding in reality. That's one reason I like to garden. Working with soil to plant seeds and grow vegetables literally grounds my electrical system.

To establish a mental grounding here, I show you how math puts a strong skeletal structure into the diaphanous I Ching oracle. That's why I emphasize that each trigram is fractal shorthand for a chaos dynamic. It grows on the p-tree, where each vp3 window combines chunky units and analog flow to establish a unique chaos dynamic in the trigram family. Bonding two trigrams into a hexagram merges their two chaos dynamics into a co-chaos dynamic.

The dp-tree paradigm lets all 8 possible trigrams bond by pairs into all 8 × 8 possible combinations as 64 hexagrams with the ability to morph into each other via changing lines. The math-minded will appreciate that the hexagrams are not only 2^6 lines = 64 stacks of binary, unitized bricks. They also develop $2^3 + 2^3 = 2^6$ lines = 64 co-chaos patterns of analog, vibrational interplay. This highlights exponential number's quixotic gift, the paradox of adding exponents to do multiplication. Result: the 64 hexagrams, each with its own unique dynamic of fine-tuned, analinear co-chaos.

In later volumes, we'll explore how co-chaos generates the 3D spacetime of our bubble above the mobic scale...and tucked below it, the 3D timespace of that other bubble. We'll examine the ubiquitous vectors of tension looping across both bubbles, merging into a tensor network that sets the arrow of time in this bubble and the arrow of space in the other bubble.

The narrow gauge of that other bubble's space arrow crush-converted its antimatter into speedy tachyonic energy that developed into a huge, unified mind. It set up the conditions in this bubble that slowly developed all of us little living beings. We are like bacteria in the gut of the Double Bubble universe.

Due to the inherent duality embedded in its origin via numbers, our universe is forever rebalancing and evolving its polarities, forever striving for the next transcendent third way that reaches a more differentiated unity, arriving again and again at higher levels of discernment, understanding, reconciliation, and appreciation. This is what we live, and we humans discuss it constantly.

We cannot be perfect or perfected. But we can be whole, for the process itself is perfect. To be whole, our process needs to include recognizing both the sunny and shadowy aspects of life and turning them somehow to constructive ends despite any negative pall. The I Ching continually suggests how to turn each difficulty toward a good end...and new beginning.

I think of the ancients who studied nesting cycles on different scales in events, who pondered dreams, who lived a philosophy of transcendent connectivity, who predicted wheeling patterns rather than specific events. Then Taoism split into various schools that went ritualist or Confucianist, sometimes degenerated into power wizardry, or became fixated on alchemy to achieve physical immortality. Modern Taoism has several branches and amalgams. For an overview, I recommend *Taoism: An Essential Guide* by Eva Wong.

"What likes to go together?" ancient Taoists used to ask. Finding it out for yourself, weighing its value, can change your own fit in the pattern...which changes everything for you. Your experience deepens as your changes alter the events that synchronize around you. At each new zoom of increased awareness, the terrain shifts; bridges appear from one part of your life to another in ways that you'd never expected. Perspectives open onto undreamed possibilities.

In Jung's phrase, this is synchronicity or "meaningful chance." He termed it "acausal" because ordinary, cause-and-effect logic cannot see a cause. However, there is a cause, but it's not linear. Rather, it's not *just* linear, but also analog...both unitized sequences and webby resonances. Analinear dynamics synchronize the spacing of matter and timing of energy in emergent events. Synchronicity beckons rather than commands; it suggests a path more than it demands. It develops networks that resonate in clusters of relationship where meaningful chance occurs as a natural part of the analog, relational function of numbers, always ready to cue subliminal nuances at the edge of awareness.

In this universe, we perceive much of our world's patterning through an intuitive, feeling-tone response to its networking relationships. We notice whatever is here because we somehow resonate to it. We are vibrating to resonances hidden in the nested patterns of number itself.

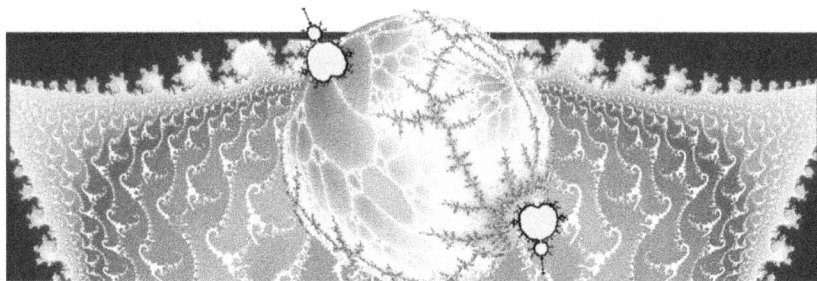

Chapter 17: I Ching Math Can Shorthand Space

1. Period-doubling/exponential growth in 3D space

The development of spatial dimensions in this upper bubble uses period-doubling and exponential growth on the path from 2 to 8. The graph below shows how both number methods use that path to move from point to line to plane to cube, thus developing the spatial shapes on the right.

Period-Doubling   *Exponential Growth*
   of Points   *of Points*

PERIOD 8: **8 points** = 2^3 = **8 points** = **3D**

CUBE

PERIOD 4: **4 points** = 2^2 = **4 points** = **2D**

PLANE

PERIOD 2: **2 points** = 2^1 = •——• **2 points** = **1D**
LINE

PERIOD 1: **1 point** = 2^0 = • **1 point** = **0D**
POINT

The 2's doubling/exponential growth builds spatial dimensions

Points in space bifurcate from Period 1's zero-point to Period 2's line, Period 4's plane, and Period 8's cube, which is where 3D space appears in our experiential reality.) Check it out. Verify for yourself that each rising level on the left has a period number that correlates to the number of points on its specific geometric shape…and also has an exponent that correlates with its number of dimensions. (Later on, I'll discuss the special case of 3-edged shapes…triangles and tetrahedrons, so wait for it….)

2. Trigrams can shorthand spatial dimensions

Three levels of p-tree growth can describe this bubble's spatial dimensions. Its yin-yang shorthand can express the dynamics of the p-tree's three levels of growth as increasingly refined polarity that culminates in 8 trigrams.

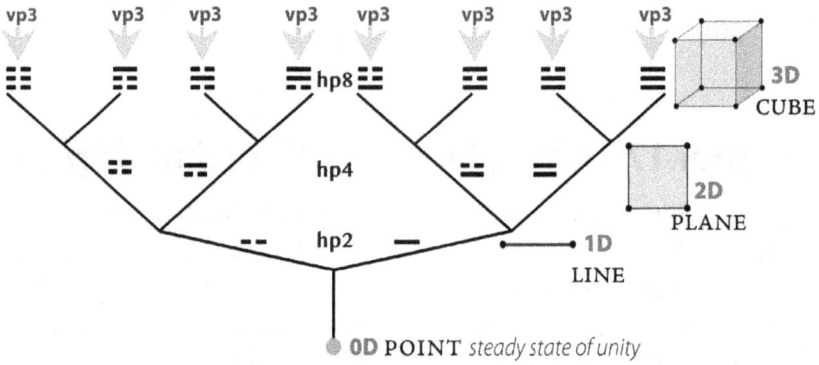

Polarized period-doubling exists in our spatial dimensions

Yin and yang describe a 1D line's spatial polarity as the first fork's two branches establish a horizontal period 2 window (hp2). The 4 bigrams describe a 2D plane's polarity at hp4. The 8 trigrams describe a 3D cube's polarity at hp8.

The 8 trigrams develop their polarity by rising in 8 vertical period 3 windows (8 vp3s). Yet the culminated polarity of all 8 trigrams appears at the third level of forking in just 1 horizontal period 8 window (1 hp8). Space's polarities operate in both kinds of windows, and both kinds are necessary to generate 3D space.

Here's how the I Ching's polarized math bonds together our bubble's spatial dimensionality. Below, start on the left at a 0D point of space. Moving rightward, the neutral 0 extends or bifurcates into a yin pole and a yang pole, which together define a 1D line...then, into 4 bigrams that define the 4 lines of a 2D square...and finally, into 8 trigrams that define the 12 lines of a 3D cube.

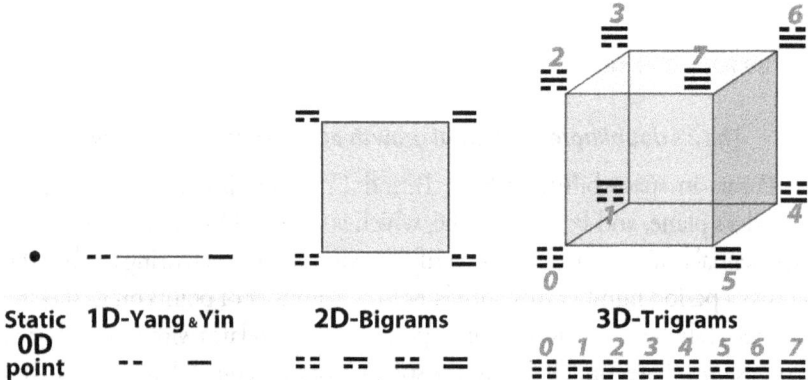

Constructing space with yin-yang polarity

You already know that space is polarized. A compass with **N**, **S**, **E**, and **W** showed you that long ago. But why is space polarized? It's because polarized pulsing at the ultra-tiny mobic scale projects invisible tension lines that create our flexing 3D spatial lattice that can be shorthanded by the I Ching math.

The graphic below represents one cube in the invisible lattice of 3D space in our upper bubble. The cube has 12 dotted-line segments that represent its 12 tension paths. Each line is tipped with a yin pole and a yang pole, for a total of 24 poles. Thus the 12 tension paths and their 24 poles establish a single cube in the 3D lattice of our upper bubble's space.

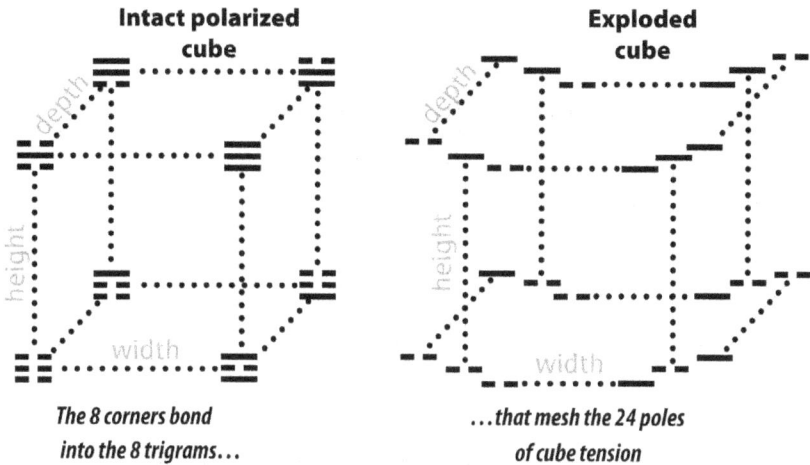

Intact polarized cube

The 8 corners bond into the 8 trigrams...

Exploded cube

...that mesh the 24 poles of cube tension

Deconstructing a spatial cube by sorting out its 24 pole

How do the poles connect? On the left is an intact cube. On the right, it is deconstructed into 12 segments tipped by 24 yin/yang poles. The cube has 8 corners. At each corner, the ends of 3 tension paths converge their 3 poles. All 24 poles in all 8 corners sort out and bond together into 8 trigrams.

A simple algorithm does the sorting to deliver a unique trigram at each corner. In each corner, the poles use this bonding algorithm:

DEPTH POLE = *TOP LINE OF A TRIGRAM*

HEIGHT POLE = *MIDDLE LINE OF A TRIGRAM*

WIDTH POLE = *BOTTOM LINE OF A TRIGRAM*

At each corner of the cube, 3 tension paths converge, and their 3 poles bond into what I call a griplet that can be shorthanded by a trigram. All 8 griplets bond together all 8 corners of the cube, which can be shorthanded by all 8 trigrams. However, due to constant scaling shifts in the space lattice, the line lengths and angles will flex and alter within their polarity bonds. A trigram can also morph into another trigram by simply changing its polarity.

**Binary Family
trigram route**

**Analog Family
trigram route**

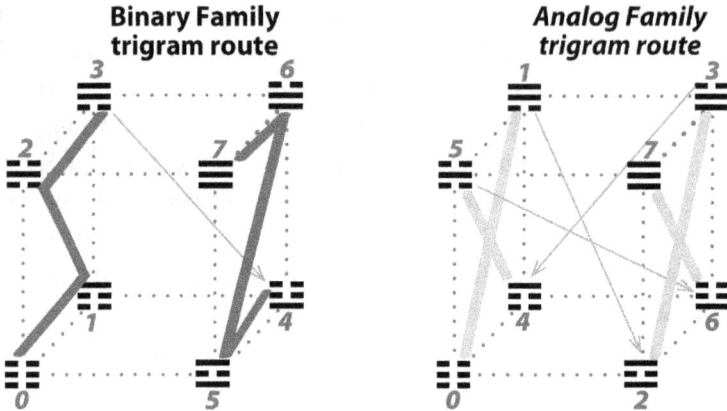

Tracking the 2 family orders on the cube

The cube on the left shows that following the binary order of trigrams sketches two Zs in depth from 0 to 7.

Binary Order

But on the right, that same cube of trigrams shows that following the analog order of trigrams sketches two Xs in depth.

Analog Order

Each trigram sitting on a corner of the cube represents a griplet of yin-yang force. All 8 trigrams at all 8 corners of the cube use both family orders to bond together this cube of 3D space. They can symbolize the griplet bonds of our bubble's 3D space lattice and also of the other bubble's 3D time lattice.

3. Hexagrams bond in the lattice of 3D space

In the scaling of the 3D space lattice, if one cube is sitting inside another, larger other, their two sets of 8 trigrams are pair-bonded into 8 hexagrams. The 8 hexagrams describe the polarized bonds of those two different-sized cubes in the latticing.

Following on the left is an image that looks like one cube sitting inside the other. It is a traditional drawing of a 4D hypercube, sometimes called a tesseract. In this purely theoretical 4D structure, three cubes and three squares are hypothesized to intersect at each edge. What? Yes!

However, this TOE says the upper bubble's 3D spatial lattice works more like the graphic to its right. Due to scaling in the lattice, larger cubes hold smaller cubes that pair-bond their trigrams into hexagrams at their cubes' corners, securing a flexible stability across the lattice.

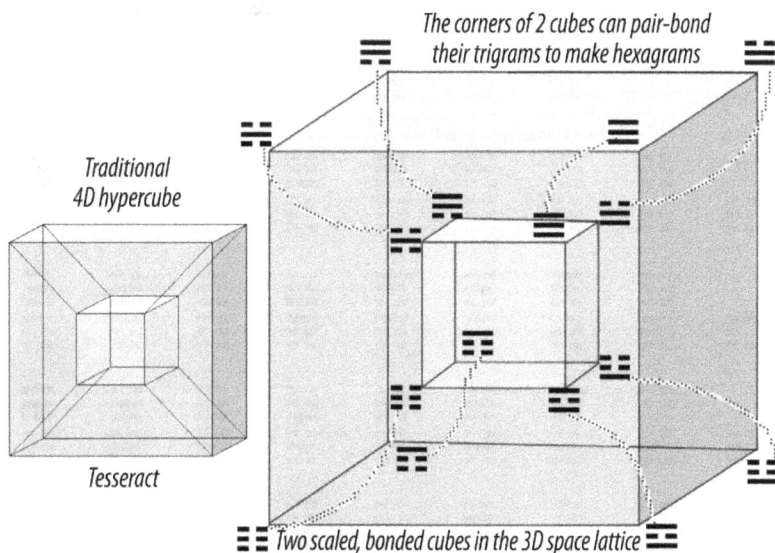

The corners of 2 cubes can pair-bond their trigrams to make hexagrams

Traditional 4D hypercube

Tesseract

Two scaled, bonded cubes in the 3D space lattice

Trigrams at the paired corners bond into hexagrams

On the right, whirl lines on the cube-within-a-cube are a visual metaphor for their trigrams pair-bonding into hexagrams. The polarity on those corners can also change and turn into other hexagrams.

4. What is the I Ching lineage?

In this bubble, gravitation is the oldest energetic force, and the I Ching is the oldest Chinese classic book. How old is it? Daniel Woolf made this cautious assessment in *A Global History of History*: "Significant Chinese thinking about the past can be traced back to ancient canonical texts such as the *Yijing*... which reached a definitive form about the end of the second millennium BC."

We don't know exactly where the I Ching began or how long an oral tradition handed it down before its math and text gained a written history. Terrien de Lacouperie, a 19th-century French orientalist, proposed that the I Ching originated in the Middle East, for he noticed it had some resemblances to Chaldean syllabaries. He also asserted that people from Akkad in Mesopotamia took the I Ching to China in 2282 BCE, led by Prince Hu-Nak-kunte, aka Yu Huang, aka the Yellow Emperor. However, I know of no recorded evidence that verifies Lacouperie's alternative history.

Russian orientalist Iulian Shchutskii dismissed Lacouperie's idea thus: "The basic ideas of Lacouperie can be summarized as follows. The *Book of Changes* is a collection of genuinely ancient materials, the understanding of which was subsequently lost and thus it was used as a divinatory text. At bottom, the book is of non-Chinese origin.....[but] this theory is based on extremely shaky 'evidence.'"

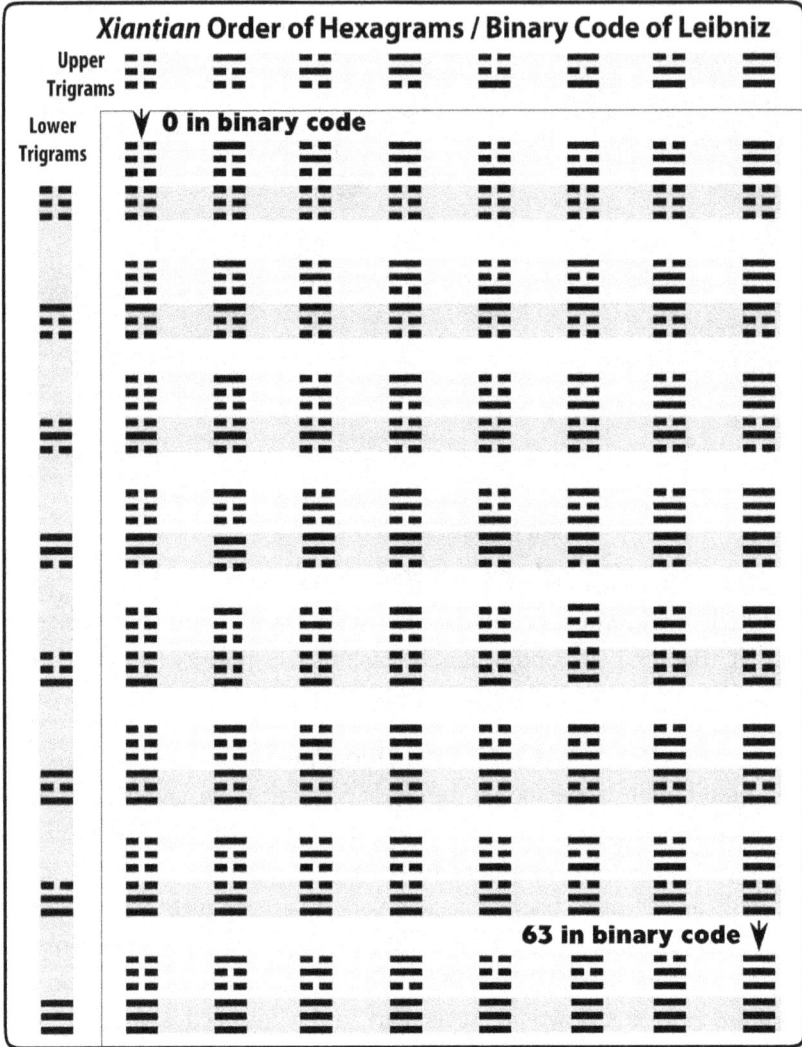

The 64 hexagrams set in xiantian order count in binary code

Whoever developed the I Ching and whatever its original purpose, the binary order holds a vast and profound mathematical regularity. It is shown here with binary 0 at the upper left corner and binary 63 at the lower right corner, reading in a Western layout from left to right.

Across the chart, each *row of lower trigrams* has just one trigram identically populated along it eight times. Each *column of upper trigrams* does likewise. Thus each column of lower trigrams merely counts down the chart in binary from 0 to 7. Meanwhile, each row of upper trigrams counts across it in binary from 0 to 7. How very orderly is this chart's binary aspect!

Across the chart's top is a lighter, horizontal key for all the *Upper Trigrams* in this chart. To the left is a darker, vertical key for all its *Lower Trigrams*. Each key counts in binary from 0 through 7. Across the chart, trigrams bond by pairs into hexagrams that count in binary sequence from 0 through 63.

This next chart takes a closer look at the top row of hexagrams. It shows their hexagram names in the Wilhelm/Baynes translation, their Chinese names, their binary and decimal number equivalents, plus each hexagram's oracle number.

Hexagram Name (Wilhelm-Baynes)	Chinese Name	Math		Oracle #	Binary #	Decimal #
RECEPTIVE EARTH	坤	☷☷	= Hexagram	2 =	000000 =	0
BREAKING AWAY	剝	☶☷	= Hexagram	23 =	000001 =	1
HOLDING TOGETHER	比	☵☷	= Hexagram	8 =	000010 =	2
CONTEMPLATION	觀	☴☷	= Hexagram	20 =	000011 =	3
ENTHUSIASM	豫	☳☷	= Hexagram	16 =	000100 =	4
EASY PROGRESS	晉	☲☷	= Hexagram	35 =	000101 =	5
GATHERING TOGETHER	萃	☱☷	= Hexagram	45 =	000110 =	6
STANDSTILL	否	☰☷	= Hexagram	12 =	000111 =	7

Here are 3 different ways to assign numbers to hexagrams

Do you wonder why there's such a discrepancy between a hexagram's oracle number, binary number, and decimal number? It's because people numbered and used Later Heaven's analog order of hexagrams to consult the I Ching oracle for about 2,000 years before Shao Yong found the Early Heaven binary order.

And perhaps you're wondering why I haven't discussed 3-sided shapes? Yes, a triangle is the quintessential 2D plane, and a tetrahedron is the quintessential 3D solid. But they did not originate in either bubble's latticing, although squares and cubes did. Triangles and tetrahedrons originated well below the quantum scale in the myriad revolving pores (mactors) of the membrane interface between both bubbles. Any 3-sided shapes that do appear in either 3D bubble's lattice are mere derivatives of a more primordial version. Volumes 5 and 6 discuss this.

Squares and cubes cannot exist in the mactors. For them to appear, dimensionality must move on out beyond the membrane's pores. How? On the revolving 2DD surface of each ultra-tiny mactor, polarized pulses (symbolized by yin and yang) develop four triangles of dimensional tension that bond into a 3DD tetrahedron with two volumes, inner 3D time and outer 3D space. But volume cannot exist in 2DD. So each tetrahedron—constantly constructing, deconstructing, and reconstructing on its mactor—projects its two volumes outward as an hourglass cell stretching far above and below that scale.

All the hourglass cells merge holographically to project our huge bubble of 3D space far above the membrane interface and a huge, mirror-twin bubble of 3D time far below it. Each hourglass cell's wasp-waist sits in a revolving pore of the membrane interface between the conjoined mirror-twin bubbles. Pondering this remarkable location is one form of meditation.

Each hourglass cell also contains three arrowing vectors that constantly 8-loop across both bubbles. As each 8-loop crosses the membrane interface between bubbles, it switches polarity from ½D time to ½D time, or vice-versa, according to that bubble's specs. All the 8-loops merge into the tensor network of an omnipresent yet single dimension that gives our upper bubble its ubiquitous ½D arrow of time and the lower bubble its ubiquitous ½D arrow of space.

Thus this universe's dimensionality was born in the mobic membrane interface between its two huge, mirror-twin bubbles. Its membrane still pulses forth the master code that generated and maintains the dimensionality of hourglass cells projecting from the mobic scale throughout the universe. In the Double Bubble, due to its pulsing master code, the bonds of dimensional latticing are extremely consistent, durable, and dependable.

All of that can happen merely because $2 + 2 + 2 + 2 = 8$. And if 2 is instead doubled and then redoubled, that also equals 8. And if 2 is squared and then cubed, that also equals 8. All three paths from 2 to 8 are not the same thing, yet they all reach, affirm, and reinforce the same goal—8—to manifest the 8 trigrams of chaos patterning. No wonder ancient China called 8 the luckiest number!

In this TOE, the master code operates as a universal genetic code. Its four primals of space, time, matter, and energy are a polarized pair of pairs, as DNA's four base molecules are a polarized pair of pairs. The master code grew on a bifurcation tree that was polarized and doubled through three levels of branches and roots to yield 64 co-chaos patterns, as did your genetic code. The master code generated this Double Bubble universe's huge body and brain, as the genetic code generated your own body and brain.

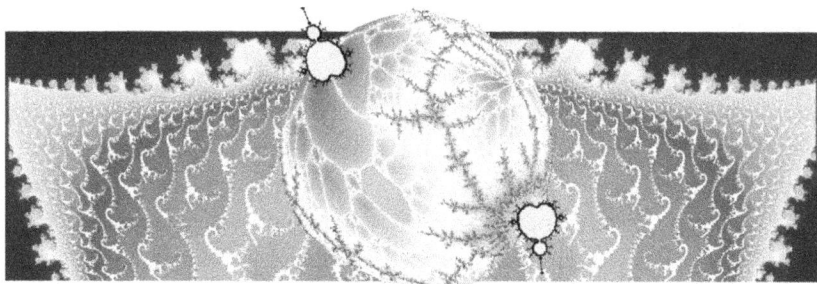

Chapter 18: A Few Tips on the I Ching Oracle

Tapping into the I Ching offers a two-way conversation with the Tao, which makes using the I Ching necessarily experiential. I can give you a few tips that may enhance your appreciation of the oracle and its answers, but only you can take the leap into experiencing it for yourself.

1. Keep a record

When you ask the oracle a question, understanding the answer and then applying it requires a lot of reality-checking. Early on, I began keeping a record of my I Ching questions and answers to check them out compared to actual events in my ongoing life. By now, I've studied more than 15,000 hexagram answers—both my own and those of clients—related to real-time events.

In Germany, I once spoke to a Chinese scholar about this experiential record I was keeping, and he said, "But that is just what the ancients must have done. They kept careful records, saw, and described it for themselves!"

With the I Ching text, the task is to pinpoint the archetypal dynamic inside its words. A verbal or mental image can tap an association that brings insight, and an analogy or picture usually comes closer than a chain of logic. *Find the image behind the words, then the dynamic behind that image.* Doing so can bring you an instantaneous flash of insight that comprehends the dynamic itself—the hexagram's fractal pattern. Only time—if you're willing to give it—will show you how an I Ching answer can spotlight the crucial dynamic in an event, past, present, or future. It can impart an unexpected insight that showcases a momentary facet on the whirling holographic jewel of your life.

This deepening trait in the I Ching is the very essence of co-chaos. It opens windows on reality that can even offer new perspectives for science. Indeed, DNA's double helix found in the 20th century is based on a fractal math paradigm recorded thousands of years ago in the ancient I Ching shorthand.

In its long tradition of written documents, Chinese thought holds an imagistic quality engineered into the very flow and juxtaposition of its brushed characters. As a result, if today you read an ancient Chinese poem in ten different English translations, you may even suppose that they are ten different yet related poems. Each translator interprets a lyrical moment of life that occurred for the poet, as indicated by a few characters. That's why the best ancient Chinese poetry is short and intensely personal, and also why the I Ching text accompanying its 64 hexagrams can get translated in so many ways.

I've studied the ancient text in China. Consulting with various scholars showed me that its written characters can be translated with wide latitude, which explains why the many I Ching translations into other languages vary so much.

However, even in ancient China, that text was obscure. Long ago, early Chinese had a written vocabulary of only about 2,500 characters. Since its written vocabulary had relatively few words, each brushed character had to carry more freight. Many characters became so weighted with multiple meanings and connotations that a message became multiply-allusive.

Nevertheless, the text's imagistic style yields a strange kind of analog accuracy. Its "truth in imagery" bespeaks a mindset that showcases an allusive, imagistic quirk in the Chinese language itself. This trait turns the mathematical shorthand of hexagrams into fractal dynamics that you can visualize. Its two-trigram math even operates visibly in the names and analogies of a hexagram.

Slowly I have developed some modern ways to explain the archetypal dynamic of each hexagram. If you are already knowledgeable about the I Ching, you'll notice that some of the hexagram titles I use here are not those familiar in the best-known English translations. Sometimes I have rephrased a title to focus on what I consider to be the core dynamic in that hexagram's message.

2. Nu Wa and Fu Xi

Apocryphal Chinese history says the I Ching began when ancient ruler Fu Xi (Fu Hsi) discovered the eight trigrams. Fu Xi was honored as the father of Chinese culture, and accounts vary on the date of his reign, ranging from 3322 to 2836 BCE. He was the first in a long line of emperor-sages.

In antiquity, the Chinese preferred their rulers to be wise rather than brawny, so Fu Xi was lauded for his superhuman cultural feats, much as Paul Bunyan was credited with superhuman physical feats in the Minnesota North Woods. Fu Xi allegedly taught the people how to write, cook, fish, trap, chart heavenly and earthly cycles…rather like a scholarly and public-spirited Paul Bunyan of the intellect.

In 1925, a silk veil and a garment painted with images of Fu Xi and his mate Nu Wa were found in a Taoist tomb at Astana Cemetery in Xinjiang, China. The couple dance in a double spiral of creation that is reminiscent of DNA's double helix. Unlike most depictions of ancient ruler-gods, these two don't flaunt scepters, weapons, money, slaves, or jewels. Instead, they hold ancient tools of measurement. Left, yin Nu Wa lifts a drafting compass to encircle yang Heaven. Right, yang Fu Xi holds a square and plumb bob to take the measure of yin Earth. Their tools bestow divine order on the spiral dance of life. China still uses the words *kuci chü* to signify *the moral standard* or trued up by "compass and square."

Nu Wa & Fu Xi in a double helix dance

3. The He Tu Plan and Luo Shu Writing

How did Fu Xi come up with the trigrams? Legend says he saw 55 spots on a "dragon-horse" rising out of the Huang He (Yellow River). The layout of those 55 dots inspired him to develop the I Ching trigrams. Below are the 55 dots of the He Tu (Yellow [River] Plan), named for the locale that inspired it.

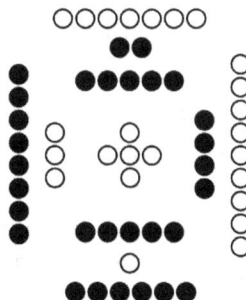

He Tu or Yellow [River] Plan

The next book, Volume 3, *Tao of Life*, correlates DNA's four base molecules seen as 55 hub or branching atoms with the He Tu's 55 black or white dots.

According to ancient Chinese history, which is hard to distinguish from legend, the I Ching's next advance came around 2200 BCE. After a great flood, Emperor Yu devised a system of controls that would provide irrigation instead of flooding. During one of his on-site inspections, a turtle climbed out of the Luo (Lo) River to reveal some intriguing spots on its shell. Its magic square of spots caused Yu to create a document called the Luo Shu, or Luo [River] Writing, named for the river that inspired it.

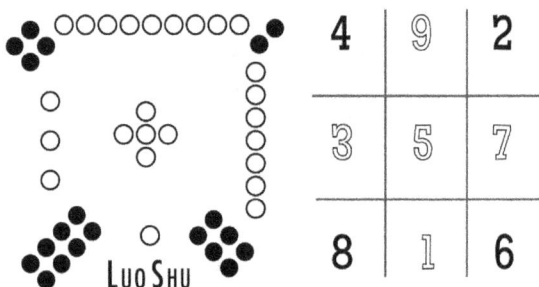

4	9	2
3	5	7
8	1	6

Luo Shu or Luo [River] Writing & Magic Square

James Legge describes the Luo Shu thus: "This writing was a scheme of the same character as the Ho map, but on the back of a tortoise, which emerged from the river Lo, and showed it to the Great Yu, when he was engaged in his celebrated work of draining off the waters of the flood, as related in the Shu."

Most likely, Emperor Fu Xi and Emperor Yu stood for legions of I Ching scholars over time, as the part stands for the whole. In ancient China, emperors were often credited with their underlings' discoveries, much in the same way that a corporate head gets public kudos for technology developed by the company employees, or professors get scholarly credit for work completed by their graduate assistants, or a nation's leader receives undue credit (or blame) for events that occurred in that country or even across the world.

Other Chinese emperors also reaped major kudos. For instance, legend says that Shennong (his name means "god of agriculture") was an emperor who started markets for exchange and barter. He introduced farm tools, agriculture, and animal husbandry. He discovered medicinal herbs and their use in the treatment of illness, so he's also considered the father of Chinese medicine.

Many small details of legendary Chinese history are beginning to be verified by recent archeological discoveries. Historians now suspect that at least some dates, events, and details of the apocryphal I Ching past may indeed be accurate. Chinese legend often does hold kernels of truth intact.

4. Getting the inner weather report

The ancient Chinese believed that consulting the I Ching oracle helped them foresee the weather of life better. I find it is true for me. However, in its metaphorical balmy day, lightning, drought, or rainfall, it does not predict any precise, quantitative specifics. Nor will it say how big a rainbow you may see afterward. The I Ching only indicates a situation's qualitative dynamic.

At night, I may do the I Ching to get my inner weather report for the next day. I often use Brian Fitzgerald's *I-Ching App* of Changes based on the yarrow stalk method. If I am away from electronics, though, I use the I Ching kit of 16 stones described in Volume 1, *Double Bubble Universe*, along with a notebook.

I'll usually receive a hexagram that changes into a second hexagram. I examine the first hexagram and its changing lines, then the morphed second hexagram. Each hexagram has its own unique flavor in the ice cream shop of reality, and any changing lines I receive are like varying sprinkles on top.

Whatever the answer may be, its hexagram math and text will provide me with an outline of the upcoming day's fractal dynamics. I must fill in for myself any unique details by living through events and using my measure of free-will choices. The advantage is that now I have a heads-up on the dynamic that is manifesting, along with some suggestions on how to handle it wisely.

Then at various times the next day, I'll stop to recall my personal weather report. Exactly how its pattern is manifesting in my daily events usually becomes evident only when I stop to consider it, since the day's flurry of activities will normally play out in scenarios so vivid that their details may sweep me up and carry me away from noticing any deeper patterns and significance.

Recognizing the dynamic that is manifesting can help me find sounder footing in a difficult situation. By identifying the pattern that is going on during a taxing event, I can respond better, sustained by an overview bigger than my anxiety or over-confidence.

It can even be curiously comforting to realize that, yes, today really is Hexagram 6 *Conflict* ☰. All day long. Strangely enough, recognizing a difficult dynamic in play can help me pour into it the best contents I can muster. It may even help me relax, not feel aggravated, sad, or guilty, and do my best. It also coaxes me to look toward the bigger picture beyond this short-term aggravation.

Tapping into the I Ching offers a way to gain insight on the fractal dynamics of my life. It lets me perceive the way of the Tao and adjust my own fit in it. Discovering that deeper pattern helps me improve the design that my life is making. If this world is a school, then I let the I Ching teach me to discern life's dynamic flow well enough to let it carry me into a better way.

5. Go moderately with the I Ching

If the I Ching is new to you, I suggest that you start with a question on something that is time-limited—a day, a specific event—so you can perceive more quickly and clearly how your answer pans out over time. And hey, do NOT overwhelm yourself by asking too many quick questions at once.

Each answer requires some meditative effort to understand its dynamic in time's hidden current. Answers about the past need internal review; answers about the future need time to watch life roll on. Too many rapid-fire questions can exhaust you with the unaccustomed mental gymnastics of trying to grasp so many metaphors, apply them, and assess their dynamics in real-time events.

Remember, this stuff is more analog than linear, more qualitative than quantitative. You cannot learn the meaning of I Ching hexagrams like you would memorize a set of multiplication tables or a list of spelling words. You need to experience each hexagram's dynamic slowly over time, keeping records as you start to discern how its co-chaos dynamic works. You can vibrate your life into beneficial change as you begin to apply the I Ching's advice.

That I Ching advice taps into your holistic right brain. It sends you over the boundary of logic into the analog zone where normal linear rules alone are not enough. The multiple resonances in an answer may even make you feel groggy or ungrounded. Moreover, if your ego becomes too threatened, it will lock down in defensive self-justification, fear, anger, and denial. If an answer feels too overwhelming or like there's too much to digest, that reaction can even drive you into full retreat from the I Ching.

The ego's linear blinders are amazingly powerful. If your logical mind dismisses your I Ching answer, just wait a day or two and look back on it again, measuring it against what has actually transpired meanwhile in your life. Often I've been amazed at my own obtuseness. My ego lives on the arrow of time, but the I Ching looks around the corner of time where ego cannot go.

So take it easy and go moderately with the I Ching at first. Apply your right brain's holistic gestalt to understand your answer, mixed with scrutiny by your left brain's steady logic, too. Then meditate on your answer at moments throughout the day to recognize its dynamic pattern.

After all, you are trying to stay in balance by doing this. Even if you may occasionally feel dizzy and disoriented with the unaccustomed mental exercise of trying to grasp an answer and apply it, practice will slowly develop your psyche's muscles. This exercise is pumping iron for your subtle body, which resides in the universal patterning long after your physical body is gone.

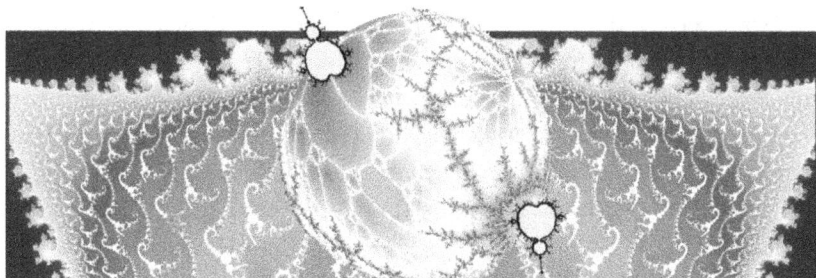

Chapter 19: Hexagram 2

1. Hexagram 2

In each volume of this series, the last chapter explores I Ching interpretation by combining logical sequencing and analog examples.

This Volume 2 examines Hexagram 2, and the interpretation is my own. Below you'll see the hexagram number, its name in Chinese and English, and its hexagram, which is a mathematical figure. After that, you will find the *Image*, *Judgment*, *Hexagram Lines*, *Line Interpretation*, *Analogy*, *Analysis*, and *Example*. First, read the *Image* and *Judgment* to get the basic dynamic of your hexagram.

Hexagram 2: 坤 *Receptive Earth* ䷁ *Co-chaos Math*

The Image

A mare with firm devotion, resolute and true,

    follows, not leads.

Sublime success by acting like humble Earth,

    not high Heaven.

The Judgment

Earth is deep with devotion,

Brings fruition without motion,

Holds without revealing it,

Knows without concealing it.

Mare-like, tireless, strong, and swift;

Cow-like in her gentle gift.

The Lines

EARTH	▬▬ ▬▬	Line 6
	▬▬ ▬▬	Line 5
	▬▬ ▬▬	Line 4
	▬▬ ▬▬	Line 3
EARTH	▬▬ ▬▬	Line 2 ☆
	▬▬ ▬▬	Line 1

Hexagram 2

☆ **denotes the most important line(s)**—*usually 2 and/or 5.*

Now overview all the hexagram's numbered lines to understand its dynamic. **Line 1** sits at the bottom of the hexagram figure. Read upward to **Line 6** at the top. Next, apply any changing lines *in your own answer* to add the nuances of their dynamics *to your own specific situation*. Add to that any insights you may find in the *Image, Judgment, Interpretation, Analogy, Analysis,* and *Example*.

Line 1 ▬ ▬
  When treading on late autumn's frost—
  Know winter's ice must soon be crossed.

Line 2 ▬ ▬ ☆
  Imitate Nature's direct and open mind.
  Cycle life without repeating it; further all without cheating it.

Line 3 ▬ ▬
  Continue unnoticed but firm on the trail.
  Serve without boasting for success without fail.

Line 4 ▬ ▬
  Narrow the opening of the sack.
  Hide one's merits from the pack.

Line 5 ▬ ▬
  Clothe the loins in yellow brilliance.
  Keep the golden mean's resilience.

Line 6 ▬ ▬
  Dragons fight in chaos chilling.
  Heaven and Earth's blood starts spilling.

2. The Line Interpretation

Next, to interpret your answer, take the general tone of your hexagram and add to it the specific influence of each changing line you received…it's like adding sprinkles atop your ice cream cone's basic flavor. If you asked about a long-term event, the answer may show a dynamic that appears repeatedly on scales large and small over a long time. Short-term issues may have answers where the dynamics are of very short duration.

Remember, a hexagram only tells you the dynamic form of the issue. You will fill in the contents of its specific details by how you handle it. Basically, the I Ching advice is a "heads up" notice that gives you a chance to make choices that can handle events more successfully, perhaps even modify the outcome.

Hexagram 1 and Hexagram 2 are both unusual regarding their changing lines. For these two hexagrams (but no other), if *every* line changes, the result is extremely beneficial. Hexagram 2 is wholly devoted to receptive yin strength symbolized by the docile mare pulling the plow. But if all the lines change from bottom to top, the result turns into the assertive yang energy of Hexagram 1, where yang's calm strength joins yin's mild docility to bring good fortune.

Line 1 ▬ ▬
Now the autumn-like chill throws a frost over events. It heralds the approach of an emotional or physical wintertime—dark, cold, and slippery with treacherous ice underfoot. Be vigilant and do not lose heart or your footing. Lighten this dark time by learning to skate over chilly events by moonlight instead of stumbling along, frightened in the frigid dark.

Line 2 ▬ ▬☆
Follow the way of *Earth*. It is foursquare—frank, broad-minded, immense in scope. It holds all with devoted acceptance. It works in endless cycles, yet it creates each moment anew, furthering all in the artless design that combines truth, stability, and profound depth.

Line 3 ▬ ▬
Right now, it is better to contain rather than to reveal one's abilities, yet maintain the goal and persevere. Serving the cause unnoticed will bring fruit in the end.

Line 4 ▬ ▬
Like a bag of treasure closed to avoid plundering, keep this situation under wraps for safety's sake. My old Chinese I Ching teacher in Guangzhou, Zhang Luanling, told me regarding this line, "I had experience of this condition. During the rule of the Gang of Four—during those 10 long years when I was banished to tend the land—all the intellectuals learned to keep their mouths shut. It was safer that way."

Line 5 ▬ ▬
This is the most favored line of the hexagram. In ancient China, yellow was the central color of balance. The yellow garment of moderation shines with a central brilliance, wrapping the reproductive area of the body in a balancing power that brings great good fortune. Following the golden mean of resilience after any setbacks brings eventual fruition at the end.

Line 6 ▬ ▬
Until this line, yin's receptive energy has successfully avoided conflict by incorporating all possibilities, but finally the struggle among competing energies can no longer be contained. Darkness that was foreshadowed in the first line is now realized in this final line, as yin's open receptivity at last gives way and turns to full aggression. The Chinese metaphor says the yellow dragon of yin Earth and the midnight blue dragon of yang Heaven fight in the wild beyond the regular limits, and both are wounded in the struggle, losing colorful blood.

3. The Associations

Placid Acceptance, Unnoticed Stability, Manifest Matter, Reticent Container, Ripening Womb, Bounteous Demeter, Receptive Earth, Power at the Background of Attention, Bearing to Fruition, Your own association with this archetype.

4. The Analogy

The dynamic of Hexagram 2 ☷ *Receptive Earth* maintains the cycling dependability of Earth. Its two trigrams are Earth ☷ deepened by more Earth ☷. This reiterating yin power presents a dynamic that manifests its potential in humble, earthly matter. Solid, tangible, an incubating womb, Earth complements the activating yang force that was described by Hexagram 1 ☰ *Assertive Heaven* in Volume 1, *Double Bubble Universe.*

Receptive Earth's strong devotion keeps it flexible and modest, while its enduring dependability allows completion without taking obvious, overt action. Its deep power gestates, mute and unconsidered below consciousness, subordinating its energy in the background of attention. It is able to admit all, accept all without prejudice as it works toward fruition. Tranquil advantage comes from being resolute and true, and from friendships with those of like mind. If this yin energy can follow devotedly in the right way, it will attract its proper leader. But if it insists on forcefully taking the foreground to lead, it loses its own profound identity and goes astray.

5. The Analysis

In Hexagram 2 ☷ *Receptive Earth*, the two trigrams of Earth depict a redoubled yin energy whose receptive power brings into humble, materializing birth the lofty conception that was already quickened by the high-flying yang energy of Hexagram 1 ☰ *Assertive Heaven*. Notice how this all-yin figure of Hexagram 2 ☷ is a photo-negative image of preceding Hexagram 1 ☰ that reverses its all-yang polarity.

To symbolize yin's nature, ancient China chose to describe the dark, firm, germinating power of earth, whose opaque matter allows itself to be plowed and sown with seeds that are then covered up and hidden away. This is the opposite of heaven's yang energy that assertively lights up any cranny that a sunbeam can penetrate. Yin recalls the solid, tangible earth holding within its chthonic richness the secret of new life burgeoning from unknown places. Yin keeps all buried in dark connection. It is a treasure house of secret possibility.

This deep, natural force of yin moves slowly and inexorably. Like the womb, its power is dark, multitudinous, unseen, yet not calculating or contrived. It does not boast its virtues in an extraverted litany of championships won, of opponents defeated, of gloats over enemies slaughtered. By its very modesty, yin manages to sustain the holistic development that brings conception to birth. Its unplumbed wisdom endures the toil of cycles repeating on scales large and small, in patterns that are similar but never exactly the same. Its effortless, seemingly tireless constancy bears fruit. Like a mother with many beloved children, yin does not judge or disown, but rather, embraces all.

This passive embrace of yin is a mysteriously accepting energy. It describes the receptivity that allows a person to be led. Through devotion and love, yin brings all things to birth. This holistic, relational energy can ripen and make manifest the potential that was sparked by Hexagram 1's yang force.

Sometimes fate brings a question to the oracle that delivers this hexagram answer. In such a situation, the wise person, aware of the great power in yin's receptive force, willingly follows the Tao to make the best use of this accepting yin dynamic, rather than choosing to fight events blindly, willfully.

A comment often attributed to Confucius says that Earth can contain all because its nature is so deep. The idea suggests that Earth is so receptive that it can hear out all versions of an event, even those unlikely or hostile, and reconcile them in a deeper truth.

If Earth is allowed to embody its modest, enduring, supportive nature, it can last a long time as it submits to the task of bringing things to fruition. This great force, said the ancient Chinese, is like a faithful mare—strong and swift as a stallion, yet also docile as a cow, willing to be yoked to the plow where a stallion would not.

Hexagram 2's all-yin lines ring with the resonances of doubling and redoubling acceptance. The yin dynamic is receptive, holistic, cyclic, and process-oriented. In this way, it complements Hexagram 1's all-yang force, which is assertive, focused, linear, and goal-oriented.

Thus yin and yang energies both have their strengths and limitations. Yang energy, when carried to an extreme, can polarize events into constant conflicts rife with yang aggression. But it is just as debilitating to agree passively without end in a too-yin manner that tries to absorb all opposition and hostility, never offering any difference of opinion, disagreement, or dispute.

Both yin and yang have their place, value, and moment, but both turn harmful if overextended and carried too far. Developing the proper timing of polarity-switching in the co-chaos patterns opens what the ancient Chinese call the path of the Tao, and what the ancient Greeks call the *kairos* of right timing.

In 1991, I was sitting in the library stacks of Jinan University in Guangzhou when I came across Arthur Whaley's *Three Ways of Thought in Ancient China*. In it, Whaley recounted a tale that seemed to be an ancient variant of my own dream about the birth of this universe.

That old tale goes like this. Confucius went to see Lao Tzu and found him basking outdoors with his hair just washed and spread out to dry in the sun. Lao Tzu sat so utterly still that his body seemed to be melted into nature, lost in time. Confucius retreated silently, careful to leave the scene undisturbed, but he returned another day to ask Lao Tzu what he'd been doing that day in the sun.

The Taoist sage replied, "I...voyaged to the World's Beginning.... I saw Yin, the Female Energy, in its motionless grandeur; I saw Yang, the Male Energy, rampant in its fiery vigor. The motionless grandeur came up out of the earth; the fiery vigor burst out from heaven. The two penetrated one another, were inextricably blended and from their union the things of the world were born."

Hearing this ancient tale, a Jungian might consider "the Female Energy, in its motionless grandeur" to be like the dependable, solid earth of this planet that is even called Earth. The two all-yin trigrams of Hexagram 2 represent it. We spring from this mysterious mother and fall back into it. Earth is the container, the womb, the tomb, the unconscious depths, the dark entrance and exit for life itself. And while we live, Mother Earth has a stable willingness to hold us; it does not judge us in a narrow, binary way. Its energy is embracing, relational, holistic, process-oriented, holding wide networks of relationships. Yin has an analog range that encompasses all possibilities. Its power can be deeply nourishing or abysmally frightening—and the full gamut in between.

Earth's yin power is a great fructifying force that can embody stability for a long period without stagnation. It is the pregnant process of endless mystery always becoming. This vast yet retiring and humble force does not lead; it follows an invisible trail. It suggests rather than commands. If yin demands to lead, it loses its way. However, when yin suggests, its vast force is applied correctly and fruitfully. The result is enduring, receptive, fertile, endless becoming.

But what happens if yin contains and contains until there's nothing left but engulfing yin acceptance with no judgment? Even the most pliant container must eventually hit its limit. The most passive acceptance is finally, paradoxically forced to become active resistance. That is the way of the Tao, and of co-chaos patterning. There must be polarity-switching. In Book 1, it happened for Hexagram 1 ☰ when so much all-yang drive just wore itself out by Line 6.

Likewise, the yin receptivity of Hexagram 2 ☷ finally becomes too all-inclusive by Line 6. Too much yin behavior becomes encompassing, smothering, and doomed to breed an eventual crisis. Containment carried beyond its sustainable limit turns into the site of turmoil, and that breaking point breeds active dragons of conflict. Hexagram 2's final line 6 portrays a chilling battle where yin and yang dragons of energy tear destructively at each other, merciless in the chaotic fray, with both of them spilling colorful blood.

6. The Example

The dynamic energy of Hexagram 2 *Receptive Earth* is familiar and significant to me, since it has colored my life to an unusual degree. Unbeknownst to me for the first 45 years of my life (not until it was identified by the I Ching's natal hexagrams) Hexagram 2's energy had already been shaping my daily life.

What are natal hexagrams? They reveal your life's four major fractal patterns. They are found by combining I Ching co-chaos math with Ming dynasty astrological math. The proper technique is discussed in *Astrology of the I Ching*, whose dense text and algorithm come from the *Ho Map Lo Map Rational Number* manuscript, translated by W. K. Chu, and edited by W. A. Sherrill.

Over 35 years, I found several Windows and Mac computer programs (now outdated) that computed natal hexagrams correctly using this algorithm. But be aware that some current websites offer what purports to be a natal hexagram reading, yet its algorithm is not authentically based on the original *Ho Map Lo Map Rational Number* algorithm, so it does not work accurately.

The correct technique divides your life into two time spans: your earlier years and later years. For each time span, you'll get a pair of hexagrams. Each pair describes the major dynamic during that time span, cycling on scales large and small. That first hexagram pair will describe the major dynamic cycling in your earlier years. The second hexagram pair shows a new major dynamic that cycles in your later years.

Each time span has an initial hexagram with one changing line that morphs it into the second, concluding hexagram. Throughout a time span, its pair of hexagrams will describe a dynamic that cycles in variations large and small. The earlier span of years tries to prepare you for what is coming in the latter span.

When does one's life shift over from cycling its early pair of hexagrams into cycling its later pair of hexagrams? That varies. To see how long an early time span holds sway, tally up the number of years indicated by its initial hexagram's lines. Each yin line counts as 6 years; each yang line counts as 9 years.

For instance, my early life cycled these two hexagrams: Hexagram 2 ䷁ *Receptive Earth* (with changing Line 1) morphing into Hexagram 24 ䷗ *Turning Point for the Better*. To find out how long my early dynamic's time span ran, tally it up: Hexagram 2 ䷁ has (6 yin lines × 6 years = 36 years) for the early life cycle, thus carrying me up through age 36 (minus a few months due to the ancient Chinese way of computing life's first year of life).

My first 36 years were described accurately by the cycling dynamic of that early pair of hexagrams: Hexagram 2 ䷁ *Receptive Earth*, changing to Hexagram 24 ䷗ *Turning Point for the Better*. Their paired dynamic continually sequenced in my life on many levels and scales during the first 36 years. I lived oddly adrift in a vulnerable, dreamy mindset during those years. I was open to empathic connections with those around me, whether I chose it or not, whether I liked it or not…silent empathy with strangers, animals, even with nature itself.

But before I ever heard of the I Ching, I already knew that some unusual degree of receptivity was operating in my psyche. What really clinched for me the accuracy of my early natal hexagrams was that single changing Line 1:

Line 1:   When treading on late autumn's frost—
Know winter's ice must soon be crossed.

Line 1's wintery metaphor describes how I often felt during those first 36 years. In that time, I sensed energies and feelings that I could not understand or explain. Sometimes they were cheerful or joyous, but more often, they were unpleasant. I sensed things about our extended family that none of us would openly acknowledge. It was daunting to me. I also sensed emotional currents in strangers around me. I found heavy vibes often lurked under an ostensibly "normal" surface, glossing over an interior turmoil, fear, anger, or insecurity.

I dreaded crowds. Even sitting in a classroom was often uncomfortable. Mostly I felt unable to tune out the discord or do anything about it. That was a chilling fate, to be tossed about so much on the waves of other people's emotions. It made me fear unknown groups and crave quiet.

Not understanding a general sorrow and discontent in the human condition or why it was impacting me so much, often I fell into gray moods and supposed I must be depressive. But my own life as a child was actually pretty good, so why was I so sensitive to the suffering of others? Why couldn't I just pretend people were happy? Especially if they pretended it, too?

Obviously, I was just too sensitive. First, I'd feel someone's pain. Then I'd become annoyed at myself for empathizing too much. I seemed thin-skinned, too tender for this life, too blue-eyed (as German slang puts it). Because of that, I did not spend much time making waves; I spent my time feeling them. In my own life, I was a reactor, not an enactor of it. How to alter that?

My great relief became reading books, and most of all, fiction, because it put into words and events the various emotions that I sensed in others around me without understanding any plotline behind what they were feeling, and having no explanation or rationale for it. (I did not know about empaths yet.)

Even better, I could find relief in nature. It gave me a break from the complexity of human emotions. Animals were far nicer to be around, and I could get along with them easily. Since their emotions were simpler and cleaner, they accommodated me on a level that humans did not.

That early life cycle was pretty dark for me. Emotionally I plumbed the dark side of yin receptivity, learning painful things that I did not want to know.

Yet I also kept hoping, and I was also continually somehow finding a turning point for the better in events, as Hexagram 24 also predicts for those early years. I eventually discovered how to maintain a calm candor, no matter what the setting. I finally became a teacher in the very classrooms that I'd once dreaded.

Then when I was 36 and teaching in college, one of my students helped me see what I'd unconsciously achieved during the time span of those difficult, too-sensitive early years. Bill was a Vietnam war veteran diagnosed with post-traumatic stress disorder. He was one of about 30 veterans who attended my college classes and lived in town as outpatients from a nearby VA hospital.

That semester, after grades came out, Bill brought his wife into my office to meet me. She held a large oil painting. It showed an old, wooden galleon tossed on dark, stormy waves, but it was lit from above by a full moon, for through a rift in the clouds, a beam of moonlight shone down on the ship.

Bill said he'd asked his wife to paint the picture for me. He said it was to express how much I'd helped him in class. I said, "I'm puzzled. How did I help you?" In my mind, I'd treated him no different from the other students.

Bill said until that class, he'd felt like a tossed ship, but my calm, humorous attitude provided a steadying beam of moonlight, and just being in the classroom all semester helped him more than what the VA Hospital did for him. He said I showed him how to live in right attitude by example, because I kept such a steady, equitable, enjoyable atmosphere going in the class.

I was stunned. His words made me realize that trying to turn my super-sensitivity about pain toward a better resolution must have worked. It certainly had paid off for me, and moreover, it apparently also paid off for someone else. I'd helped this vet to stabilize after wartime, at least according to him.

Then in the next semester, on a Monday, another PTSD veteran outpatient stopped by my desk after class. He told me that over the past weekend, he took the shotgun out of his mouth and his toe off the trigger because he decided that he'd rather come to class this week. It would be more fun. Nonplussed, I said, "Good." We talked a bit. Then I said, "Let's walk over to the counseling office."

Wow. Maybe all those dark, chilly early years of my life were actually in aid of something? At age 36, I'd not yet learned about natal hexagrams, nor even about the I Ching, nor knowing that those chill, early years were depicted by changing Line 1 in Hexagram 2. But fortunately, since Hexagram 2 was all yin lines, each line's dynamic lasted only 6 years, and so I was able to move relatively quickly on through those two hexagrams of my early time span.

Perhaps I needed to experience cold, dark Line 1 to prepare me for what would come later. Always your early dynamic cycle forges the tools that you'll need during the later cycle, and my later life carries a dynamic of fulsome sweetness. It begins with Hexagram 16 ☷☶ *Enthusiasm.* A single changing Line 4 reverts it into Hexagram 2 ☷☷ *Receptive Earth* again, as my fourth and final hexagram. So for me, Hexagram 2 ☷☷ *Receptive Earth* is both the first and the last of my four natal hexagrams. Thus, out of all 36 lines in all four natal hexagrams, 34 of my lines are yin. For good or ill, that's a lot of yin!

To see how long my latter cycle ran, tally it up: Hexagram 16 ☷☶ signifies: (5 yin lines × 6 years) + (1 yang line × 9 years) = a total of 39 years for my later life cycle, running up through age 75. Later years become more self-directed.

Now at age 79, I can verify that my four natal hexagrams in the earlier and later time spans reveal four major fractal patterns that shaped my life. I rejoice now that Hexagram 16 *Enthusiasm* describes so well the latter span. Synchronistically, the root of *enthusiasm* is Greek *theos* (god). My days are filled with awe, delight, and devotion. I'm awed by the divine gift of life, I delight in the fabric of life itself, and I'm devoted to its creative source—the cosmic, purposeful connectivity in the grand organizing design that I call God for short.

I can attest to Hexagram 16's description of being aware of large harmonies and rhythms in nature. Witness my interest in the co-chaos patterns that reside in the universe, and in our own human natures. Hexagram 16's single changing Line 4 also brings me a blessing. From the Wilhelm/Baynes English translation:

Line 4: The source of enthusiasm. He achieves great things.
Doubt not. You gather friends around you
As a hair clasp gathers the hair.

Line 4 accurately describes how I love to gather ideas and people into relationship, much as a clasp gathers hair. And due to this series, it also indicates my widening circle of friends and acquaintances in other countries.

When I finally learned my natal hexagrams, I realized it was up to me to optimize their major trends in my life already cycling in fuzzy fractal outline, to fill in their details the best I can by my own free will choices. I deal with those fractal forms in the same way that I deal with my physical attributes of hair, skin, eye color, gender, build, age, etc. I acknowledge that and make the best of it.

You see, I cannot escape the basic dynamics in my natal hexagrams, but I can handle them well or badly according to how I fill in the emerging unique details by using my own attitude and free-will choices. Using the I Ching to get the daily weather report for its emergent dynamic also helps me handle myself in a more objective way, acknowledging my strengths and weak spots.

The isolating chill in that dark, frustrating, empathic fog of my early years became the gloriously resonant harmonies of connectivity in my later years. I eventually married the most cheerful, affable, astutely pragmatic man I'd ever known. It was a pleasure to live with him and share our days while we could.

My whole lifespan is filled with yin energy. I first explored the dark side of yin, but now I know the deep riches that yin can experience by opening up to the divine. Wherever my life has taken me and whatever its ongoing outcome, I often find myself smiling and saying, "Yes, I accept."

7. Following the Tao

It has taken me more than 30 years now to develop this series, partly because for a long time, I didn't really know where I was going or how to get there. I only knew I needed to find a way to describe what I saw about our wonderful universe in that great God dream. I just kept following my nose because it seemed to sniff out which way to go. As a symbol for instinct, the nose knows.

But for the longest time, the rest of me did not. I did not understand what I was writing or know what it would look like. It appeared to be only the seed of an idea at first. Growing it at all meant I had to develop at least a modicum of knowledge and vocabulary in scientific and mathematical fields...areas where I'd never paid much attention before. Arrrrgh!

I soon began to write down the first awkwardly worded concepts, but I had to shield the muddled, developing manuscript from my own critical ire. I had to treat it as if it were an infant incapable of doing much except be carried for the longest time before it could even begin to toddle around.

Slowly, gradually, I learned just enough math, genetics, and physics to keep going a bit further, finding more vocabulary to keep filling in the evolving details of this TOE. Like an embryo gestating in the womb, its text kept morphing at each new stage of growth. What? It's developing limbs now? Those clarifying features, that refinement of details? Wow, who knew? What comes next?

My inchoate inner vision kept opening up into new areas of scientific theory—genetics, chaos theory, cosmology, quantum physics—but it all developed in such slow stages, always gradually coming together just enough to become just coherent enough for me to continue formulating this series.

I apologize for all the lacks in its development. Even within a single book, each part of the TOE concept so hinges upon the rest that you may find me lagging over a detail too long, or leapfrogging back and forth as I try to bring a topic up to speed with the rest of it. But it hopefully all leaps together finally.

Fortunately, within 3 months of that dream where God returned, I went to an I Ching workshop at Austin's East-West Center, and somehow I realized this math might shorthand the master code I'd seen operating at the universal level in my dream. Within another year, I began to understand the polarized bifurcation tree, and then the double p-tree that ratified my strange dream of candelabra mirrored on a tabletop. Slowly I could comprehend more verbally, rationally, consciously what I had already grokked wordlessly inside that dream.

To spend enough time on this series and flesh it out, I also imposed on myself a lot of *yes* and *no*. "No! Do not watch too much TV. Do not go out to a long, talkative lunch too often. Yes! Sit down and write. Rewrite. Make it clearer. Forget the candy bars, the carbonated drinks. Eat properly so that ill health doesn't slow the progress. It's slow enough already."

"Don't spend too much time with noisy people who like to live in a constant melodrama of raging histrionics. Watch a movie instead. It's cheaper, better done, and resolved more quickly. Remember to garden. Touch animals. Hug friends. Look at the stars twinkling through the midnight branches. Look at my shadow and reconcile it. Sleep in peace. All of this helps manifest the work."

Why bother? Because I want to show you the master code that created our universe, the tree that it grows on, producing the same co-chaos patterns that are embedded in your DNA, in mine. Its branches stretch above into cosmology, and its roots probe down even below quantum physics.

Both those outer realms are far beyond our everyday senses, so science has set out to explore them with Voyager telescopes and quantum microscopes. Yet even then, we still cannot look beyond space itself or even into the quantum level of physical reality with any tools or techniques yet devised in science.

Therefore I have sought to do it by using something beyond physics… namely, philosophy in its metaphysical branch. This is one woman's effort to chart a physical reality extending beyond the access of our physical senses and the mechanical sense extenders of our telescopes, microscopes, and computers.

How tiny you and I are in this universe we call home! Yet each of us occupies a unique moment in its space and time, its matter and energy, its mind. By fulfilling your own destiny, tangled and knotted though it may seem at times, you manifest the pattern of yourself in our universe. You do it by cultivating and refining your mind, your feelings, your behavior, your values, your goals. You can slowly discover the shape of your own unique reason for being here, fulfill and beautify its design woven in the intangible fabric of your own soul.

By discerning and working with the dynamic flow of the Tao instead of fighting against it or drowning in it, you enhance the quality of your own life while furthering the universe's distant and unknowable aims. The universe wants you to succeed. It has embedded the flow of the Tao into everything so that you can catch its drift and ride it where you need to go.

8. A friendship that came through the I Ching

Here in Volume 2, this final chapter likens the dynamic of Hexagram 2 *Receptive Earth* to the purposeful, gentle devotion of a mare being guided to plow the field that will reap a harvest. However, each final chapter in this series also holds an alternative way to view the hexagram under examination, as seen from the perspective of a longtime explorer of the I Ching in art and meditation, Adele Aldridge.

In 2003, I needed an I Ching font for my Apple computer, and my homemade attempt looked tacky. An internet search led me to Adele Aldridge's website, with a page selling the I Ching font she had made. I ordered that font and paid for it, yet due to some internet quirk, no window showed up for me to download the font. I waited a day for a product email, and then I sent a follow-up email requesting the font.

Instead of just sending me a link to the font, Adele accompanied it with an interesting personal email of apology. It included her phone number, so I called her and found Adele delightful. What had started out as annoyance at an internet flub turned into a friendship. Not receiving that font allowed us to start sharing each other's company in exploring the I Ching.

To my surprise, about 10 years later, I found the story of our meeting in *The Synchronicity Highway* by Trish and Rob MacGregor. It turned out that in the interim, Adele had told a friend of hers, Trish, about how she and I had met. Trish eventually put the story into a book on synchronicity.

Here is Trish's account: "The Internet facilitates connections among people and often acts as a vehicle for psychic phenomena and synchronicity. In fact, if you use your computer a lot, you've probably experienced Facebook or e-mail telepathy as Adele Aldridge did.

"Adele, an artist and author who writes about the I Ching, a Chinese divination system, was writing a letter to an author when she received an e-mail related to her art business. The correspondent had ordered an I Ching font Adele had created, and wondered why she hadn't received it yet.

"Adele stared in disbelief at her monitor. The e-mail was from Katya Walter, the author of *Tao of Chaos*, a book about the I Ching and the genetic code. Not only was that book on the corner of her desk, but Katya was the author to whom Adele was writing, to ask if she would write a foreword to her proposed book, *I Ching Meditations*.

"'I'd placed the book on the corner of my desk to remind me to write to Katya. I kept putting that task off, not only because I was immersed in writing the proposal, but because I felt shy about approaching such a knowledgeable author.'"

"Adele immediately wrote to Katya and sent her the font. Katya was so pleased to hear from her directly that she called Adele and they've been friends ever since.

"This kind of telepathic synchronicity is so simple and straightforward it's as if the universe was saluting Adele for being in the flow. So the next time you experience something like this, pause for a moment and appreciate just how many details had to come together for the synchronicity to occur."

Yes, Adele and I recognized the kiss of synchronicity when it came. We both realize the I Ching speaks to each person differently, uniquely, even as it holds the whole world in its grasp. We also know that when you get an I Ching answer, it shows the co-chaos dynamic that is iterating in your situation. Once you recognize its pattern, your solution begins to emerge...if you'll look for it. How you handle that dynamic will fill in its unique details and outcome.

9. Adele Aldridge: Hexagram 2, Line 4

Hexagram 2, Line 4

Hexagram 2
RECEPTIVE
EARTH

Protect the unborn, a "tied up sack"
Carefully hiding, not yet ready to live in the world.

Sometimes an image can give you a gut feel for a hexagram's dynamic that is deeper than words can manage to do. This illustration by Adele Aldridge may help you sense the dynamic in Line 4 of Hexagram 2 *Receptive Earth*. I chose to show you Adele's illustration for Line 4 because that's how I felt about this series while gestating it, laboring to write it, experiencing how long it took to deliver.

The I Ching is one of the oldest books in literature. When I was learning to use it, I often opened four or five different books, comparing one text to another like various friends discussing the hexagram from their slightly different viewpoints. I sought what resonated most in it all for me and my particular question. That's how I learned to understand the I Ching's analogies.

The internet is a modern treasure trove of information. Hilary Barrett's I Ching study site at onlineclarity.co.uk lets you improve your understanding of the I Ching and share questions. I also recommend her book, *I Ching* by Hilary Barrett. It contains a succinct I Ching translation of each hexagram. It handles I Ching imagery, history, and key concepts well. However, rather than employing the coin method described in that book, I suggest that you instead use the 16 system for consulting the oracle itself.

You can find free I Ching programs and thousands of years of relevant texts on the web. The ebooks in this series provide links to many sites. To give the print book reader at least one online option, my current favorite text is at *YellowBridge.com.* You'll find a wide diversity of approaches to the I Ching. For example, Roger Sessions, Benedictine Oblate, wrote *Wisdom's Way: The Christian I Ching.* The I Ching is so old, so wise, so encompassing that it can embrace many approaches from different ages and cultures. I invite you to explore its historical depth and inclusive breadth without end.

FOR MORE CONNECTIONS, GO TO…
katyawalter.com
TO VISIT KATYA WALTER'S YOUTUBE CHANNEL, GO TO…
Katya Walter on YouTube

To continue this journey, read the next book…
Volume 3–*Tao of Life: The Fractal Gift*

Series Summary

1. What is our universe?

This TOE says we live in the Double Bubble universe. Its two bubbles are conjoined, symbiotic mirror-twins with reciprocal properties of space, time, matter, and energy. Science sees our white-hole bubble above the *quantum* scale, where matter and energy emerge. It does not see a black-hole bubble conjoining our bubble at the far-tinier *mobic* scale where space and time emerge.

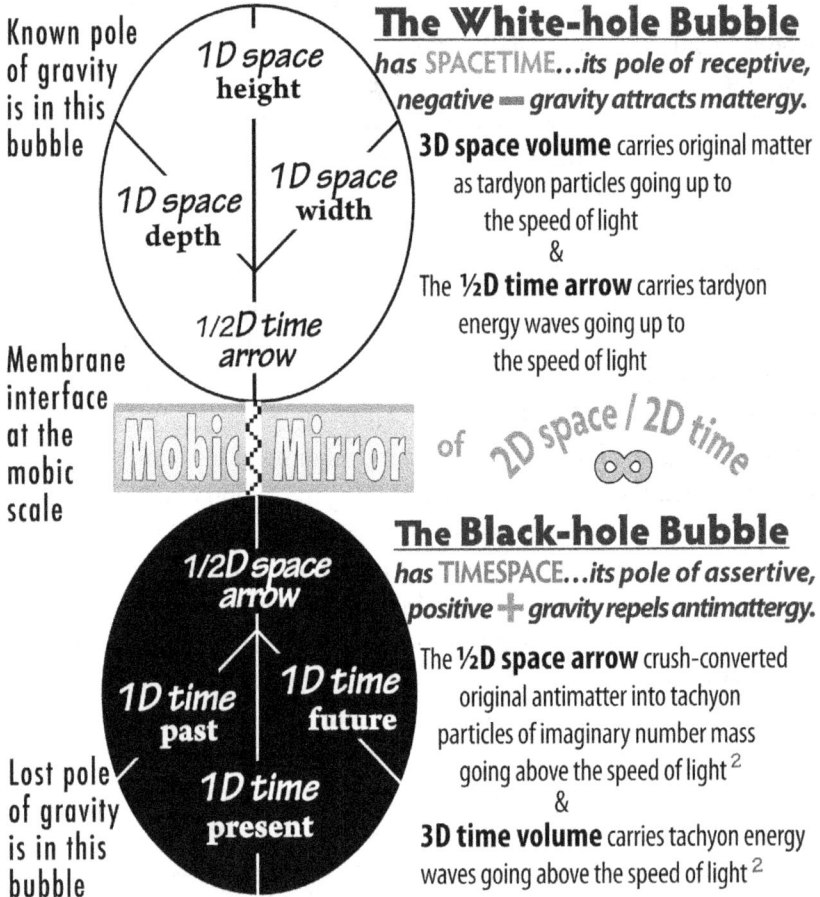

Known pole of gravity is in this bubble

1D space height

1D space depth

1D space width

1/2D time arrow

The White-hole Bubble

has SPACETIME*...its pole of receptive, negative* ▬ *gravity attracts mattery.*

3D space volume carries original matter as tardyon particles going up to the speed of light
&
The **½D time arrow** carries tardyon energy waves going up to the speed of light

Membrane interface at the mobic scale

Mobic Mirror *of* 2D space / 2D time ∞

1/2D space arrow

1D time past

1D time future

1D time present

Lost pole of gravity is in this bubble

The Black-hole Bubble

has TIMESPACE*...its pole of assertive, positive* ✚ *gravity repels antimattery.*

The **½D space arrow** crush-converted original antimatter into tachyon particles of imaginary number mass going above the speed of light [2]
&
3D time volume carries tachyon energy waves going above the speed of light [2]

The Double Bubble universe has 11 dimensions

Our upper bubble has the *spacetime trident* of contiguous 3D space with a one-way, ½D arrow of time, plus one pole of gravitation and the original matter. Its 3D space holds many material 3D structures morphing on the arrow of time, but its particle-waves are slowed down to the speed of light.

The lower bubble has the *timespace trident* of 3D time with a one-way, ½D arrow of space, plus gravitation's "lost" pole, plus the original "lost" antimatter that was long ago crush-converted by that lower bubble's meager ½D space into tachyon particle-waves going above the speed of light[2]. It powers a huge, unified mind constellated in vast 3D energy patterns in the lower bubble's 3D time.

The mirror-twin bubbles conjoin at a membrane interface of ultra-tiny, mobic pores called *mactors*. Each pore combines traits of a Mobius band and a Lorenz attractor, hence the name of *mactor* for its dynamic at that ultra-tiny scale.

How did space and time begin? The cosmegg set a dimensionless point with an *on*-pulse of being. A second *on*-pulse sketched a 1DD line of polarized tension with two poles: space and time. A third *on*-pulse set a 2DD triangle with two polarized faces: 2D space and 2D time. Polarized tension ran around both faces on a χ path much like an infinity ∞-loop.

Just one more *on*-pulse turned that 2DD-triangle into a 3DD tetrahedron. It had two volumes: 3D space and 3D time. The outer volume of 3D space projected far above the mobic scale, while the inner volume of 3D time projected far below it. Together they made an hourglass cell. That single cell replicated many times. All cells merged into our holographic bubble of 3D space above the mobic scale; below it, into a holographic bubble of 3D time.

The Double Bubble hologram merged all its ∞-loops and projected them, so now they 8-loop across both bubbles, switching polarity as they cross the interface, creating the tensor network of a single yet ubiquitous dimension made of two ½D arrows moving on an endless, polarized 8-loop across both bubbles.

The upper and lower halves of this 8-looping tensor network are polarized per bubble as either ½D time or ½D space. We have the time pole. We experience it as the point of constant *now* moving forward on the arrow of time. But that other bubble has the constant point of *here* moving backward on its arrow of space. *Here* is the only location possible in that other bubble. (There is no *there* there.)

2. Count the dimensions of this kleiniverse

Our universe has how many dimensions? Count 3.5 dimensions per bubble, making 7 dimensions in both bubbles. Count 4 more dimensions at the mobic scale itself, where every mactor's mobic warp generates 2DD triangles with polarized faces of 2D space and 2D time. This totals 11 dimensions in a layout of complementary space and time that is symmetrical across both bubbles.

The dynamic of the Double Bubble recalls a Lorenz attractor. Its two domains are the upper and lower bubbles. Its 3D space and 3D time act as three coupled Ordinary Differential Equations (ODEs) iterating along the arrows of ½D time and ½D space to evolve the nonlinear solution of reality emerging in both bubbles. The reciprocal laws of physics and the reciprocal scaling of space and time, matter and energy let both bubbles fit inside each other as a *kleiniverse*.

3. The master code uses four primals

This TOE says our universe is a huge, living organism whose fractal structure is generated by a co-chaos paradigm that iterates in self-similar patterns on many scales. Its master code uses four primals: space, time, matter, and energy. This polarized pair of pairs sort into two *carriers*: space and time…and two *cargoes*: matter and energy, polarized such that space carries matter, and time carries energy.

↓ Carrier Pair	4 Primals	↓ Cargo Pair
1. Space	←··· CARRIES ···→	3. Matter
2. Time	←··· CARRIES ···→	4. Energy

Panel: the 4 primals are a polarized pair of pairs

Our 3D space bubble has tardyonic particle-waves in vast material structures of self-similar, evolving 3D patterns. The 3D time bubble has speedy tachyonic particle-waves in vast energetic constellations of self-similar, evolving 3D patterns. In both bubbles, intricate detailing on many scales recalls the Mandelbrot set.

Both bubbles cooperate to refresh their space-time forms and update their matter-energy cargoes at a rate that makes our holographic universe appear to be smoothly continuous to our senses and mechanical tools above the quantum scale, the smallest scale known to current physics. In the universal body, old configurations decay and new ones develop. We tiny organisms in the upper bubble experience this flux as the emergent events of ongoing reality.

4. The genetic code is a variant of the master code

How can we decipher the master code that iterates the universe? We can study a lesser variant, the familiar genetic code that iterates us. It offer us some clues.

CLUE: DNA uses four base molecules: **T**hymine, **C**ytosine, **A**denine, and **G**uanine. Its polarized pair of pairs sort into two *pyrimidines*: **T** and **C**—and two *purines*: **A** and **G**. They are polarized such that **T** bonds with **A**, and **C** bonds with **G**.

CLUE: The four base molecules can pair-bond by triplets (*codons*) to make 8 × 8 = 64 molecular 6-packs on the double helix to iterate and maintain the bodies of all evolving species. Old configurations of individual organisms decay, and new ones develop. We experience this flux as the emergent lives of ongoing species.

↓ Pyrimidine Pair	4 Molecules	↓ Purine Pair
1. T	←··· BONDS WITH ···→	3. A
2. C	←··· BONDS WITH ···→	4. G

Panel: the 4 DNA molecules are a polarized pair of pairs

5. Our Rosetta Stone: I Ching, genetic code, master code!

This TOE says the ancient I Ching hexagrams of China offer a math shorthand for this paradigm. It grows on a bifurcation tree that is both doubled and polarized, i.e., it is a *dp-tree*. The I Ching's easy math can shorthand the genetic code, and their kinship gives us a Rosetta Stone with three scripts, two known and one unknown. The two known codes can help us decipher the unknown master code.

CLUE: Like DNA, I Ching math figures are also a polarized pair of pairs. They sort into two stable bigrams and two unstable bigrams: stable yin ☷, stable yang ☰, changing yin ☳, and changing yang ☴.

Yin-based ↓	4 Bigrams	Yang-based ↓
stable yin 1. ☷	←··· *STABLE PAIR* ···→	stable yang 3. ☰
changing yin 2. ☳	←··· *CHANGING PAIR* ···→	changing yang 4. ☴

Panel: the 4 bigrams are a polarized pair of pairs

CLUE: The I Ching math develops on a dp-tree, and it can shorthand the other two code variants. Below, the dp-tree has three levels of polarized forking above and below a neutral 0 seed in the middle. On the dp-tree, *minus* ⚊ stands for a yin ⚋ ⚋ fork. *Plus* + stands for a yang ⚊ fork. Its first level of forking outward develops - *yin* and + *yang* poles. The second level outward develops the four *bigrams*. The third level outward develops eight *trigrams* above and below.

Each trigram is a *vertical* period 3 window (*vp3*) defining a chaos pattern. This postulates a vital variant on Yorke and Li's *horizontal* period 3 window (*hp3*) that defines a chaos pattern in their seminal paper *Period Three Implies Chaos* (1975). In each vp3, addition by 2s, period-doubling, and exponential power together create a nonlinear chaos process so special that I call it *analinear*.

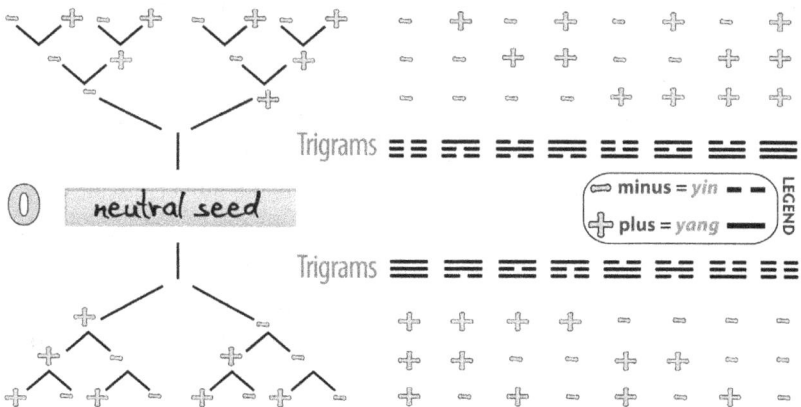

The dp-tree has 8 × 8 polarized vp3s = 64 co-chaos patterns

CLUE: Each trigram's math describes a *chaos pattern*. The dp-tree can pair-bond its trigrams into 8 × 8 = 64 *hexagram* 6-packs of *co-chaos patterns*.

Our Rosetta Stone's triple play features the familiar genetic code, the ancient I Ching, and an unknown master code. The first two codes have some shared traits that will help us discern features of the master code at the mobic scale, where polarized pulses organize into triplets of information in myriad mactors. The triplets then pair-bond into 8 × 8 = 64 co-chaos patterns that develop our universe's emergent properties. They project the Double Bubble's space-time skeleton and flesh out its matter-energy body to evolve its huge, ongoing life.

6. What are we?

This TOE says our Double Bubble universe lives, and we are like microbes living in its gut, oblivious to its larger aims. In our white-hole bubble of 3D space, we tiny, diverse, walkabout minds are powered by particle-waves of slow tardyon energy. But the black-hole bubble of 3D time holds a single, giant, unified mind that is powered by zippy tachyon energy moving at more than lightspeed[2].

Many tiny bodies with portable minds inhabit the upper bubble. But when a mind is released from ego identification by sleep, trance, meditation, or other means, it can tap into aspects of that unified mind in the tachyonic cloudbank of 3D time (some call it God or Mother Nature) processing the data of concurrent past, present, and future. That unified mind evolves its huge, beautiful, and diversified universal body. For instance, in the upper bubble, it established the far-flung galaxies and tiny micro-organisms under rocks that hold our attention.

The universal mind even delivers dreams to us nightly, in dramas that address our specific needs, fears, and hopes…but most of us have forgotten how to translate its symbolic lingo. Relearning it can cultivate a sixth sense, an ability to tap into nature's basic patterns by deep-see diving. We can access info in the tachyonic cloud of the lower bubble via shared intention and resonance. It recognizes and responds to whatever is in you—so go carefully and with good intentions. Treated wisely, it can heal and unify us, body and soul, layer by layer.

The minds in both bubbles contribute in various ways to the thrust of universal existence, which has a greater purpose beyond our own human preoccupations. Our universe plans a wider future for us as we become more conscious of our place in the whole. It has been patiently cultivating its universal life, including us among its myriad forms, hoping to evolve us enough to recognize that it too lives…and further, coaxing us to divine that there is something even greater beyond. We have the chance to acknowledge, share, and improve this destiny.

Blurb and Reviews

Co-Chaos Patterns
THE I CHING FRACTAL

Katya Walter, PhD

BIOGRAPHY

Katya Walter has a Ph.D. with an interdisciplinary emphasis from the University of Texas at Austin. She spent 5 years of post-doctoral study at the Jung Institute of Zurich, and a year of post-doctoral study in China. Dr. Walter taught in colleges and universities in the USA and abroad for 16 years before focusing on writing and lecturing. She has given numerous workshops on the I Ching, chaos theory, synchronicity, and dreams in the United States and Europe.

-:::-

FROM THE EDITOR

This book is Volume 2 in the dazzling *Touching God's TOE* series, 4th edition. This series shows how the I Ching and DNA are two known variants of a hidden master code that generated the universe. Examining parallels in all three codes offers us a Rosetta Stone to decipher the master code.

Called in Germany a "philosopher queen of the global village," Katya Walter, Ph.D., in this volume sets out the fractal foundation for the master code. She shows how it began not only as polarized binary units $(0, 1)$, but also as polarized analogs bifurcating from the zero of nothing $(0 = -1$ and $+1)$. She explores how co-chaos patterns in polarized pulsing generated our universe. Along with scientific concepts, she also examines philosophical parallels and recounts events inspiring the series.

This series began as one volume, *Chaosforschung*, published in German in 1992, and in English as *Tao of Chaos* in 1994. It was later split and amplified into Volumes 2 and 3 of the series. Its odd-numbered chapters $(1, 3, 5...)$ are more science-based. Even-numbered chapters $(2, 4, 6...)$ are more philosophical and personal.

This book has 19 chapters and a *Series Summary*, comprising 105 sections. It has a *Bibliography* and *Reviews*. It has 70 listed images, graphics, and charts. The ebook version also has an interactive table of contents and 87 e-links that act as informative footnotes. Its text is completely searchable and receives electronic updates. It is also hand-edited to hold color graphics that allow greater distinctions in images and charts. Consider getting both the print and ebook versions for a greater range of information and versatility.

-:::-

Praise For The *Touching God's Toe Series*
-:≡:-

What an interesting and inspiring writer...interesting scientifically and inspiring metaphysically! I have traveled widely, but never on a roller coaster of dimensionality before! It makes Flatland look—well, flat. Quantum organics reveals how space, time, matter, and energy mirror aspects of our DNA. And the author's take on what animals think is shockingly possible! You'll never regret picking up this series and reading it. It will take your mind to new places, and it will lift your soul along the way.

Lynn Hayden
Consultant, Singapore Institute of Management

"I find the *Double Bubble Universe* the most promising of all the TOEs being proposed currently. It involves a new topological model spanning all levels of reality and "deep-see diving" into fractal pattern recognition. It answers far-reaching questions such as 'How did our universe begin?' and 'How are telepathy and remote viewing possible?' This model deserves careful reading by the best minds of our time."

Oliver Markley, Ph.D.
Professor Emeritus, Human Sciences & Future Studies
-:≡:-

"Are you smarter than a fifth grader?" Better yet, can you bring the clarity of a child's fresh perspective to a Theory of Everything (TOE) that reinterprets standard physics data to reveal a stunningly new and elegantly symmetrical model? If you can, then this book's for you.

Dr. Katya Walter shakes the foundations of currently accepted concepts about physics, metaphysics, and the nature of consciousness. She offers a comprehensive exploration of the way fractal chaos theory forms the underlying structural dynamic that creates and allows for the ongoing evolution of both mind and matter. She also addresses the relationship of mind and matter - physically, spiritually, and philosophically - in ways not previously presented elsewhere.

This seems like a good place to mention that no weeping, wailing and gnashing of teeth are required when reading this book...it is well written in a way that is comprehensible to a general audience as well as for scientists.

If you can set aside any skepticism and/or preconceived notions long enough to allow lucid consideration of the concepts she proffers here, you may be the first on your block to recognize her Theory of Everything as the dawn of the brightest new paradigm since Newton.

Brenda Kennedy
Reader
-:≡:-

I can't decide if this is fact, sci-fi, or psi-phy. Whatever, it is truly fascinating. A brain gym of possibilities!

Frank Patterson
Aerospace Engineer
-:≡:-

My guess is that your natural reader would be a non-scientist who wants to put science and philosophy together in a coherent mental image.

David Booth, Ph.D.
Mathematician, Inventor
-:::-

I cannot emphasize enough how much I love this book. It makes the most current information about quantum physics into a conversation that can span the thinking styles of both scientists and spiritists. Katya is a dedicated dreamer, and a receiver of concrete knowledge in frontier quantum physics. There should be no separation of physics and metaphysics. There should be fluency and grace and relation to both subjects. This book achieves an understandable explanation of our human experience of dimensionality, and of our fractal nature. She proposes a new Theory of Everything (T.O.E.) If you want the most original elegant synopsis of our existence, which uncovers the mysterious forces of nature (including gravity) and therefore our consciousness, buy this book. You will have an "Aha" moment, and then you will be with me, saying, "Every life-student should be so lucky to have been exposed to Katya Walter. Reading *Double Bubble Universe* is like being in Einstein's living room."

Jennifer J. Colbert
Reader
-:::-

Simply brilliant! And I mean that literally. The clearest explanations are the least complex, and Dr. Walter has managed to take ideas from advanced physics, express them simply, then turn around and analyze the physics to present a clear, simple, and straightforward new paradigm for how the universe works. This is the simplest physics book I've ever read, because of Walter's brilliant use of language that makes these complex concepts entirely understandable. The interweaving of her 'journey-to-the-aha' adds a profound metaphysical understanding of how our universe works from the inside out. You won't regret buying this book.

Anne Beversdorf
Reader
-:::-

The author of this extraordinary book has a rare combination of qualities: an astonishing depth of vision and a genuine modesty. *Double Bubble Universe*... is exploring Katya Walter's theory of everything (TOE). A TOE is the Holy Grail of modern physics. A theory that reconciles the billiard-ball predictability of Newton's Laws with the mysterious goings-on at the Quantum level.

Dr. Walter's book proposes that Physics is blind to another domain in the universe which she describes clearly and patiently with easy-to-grasp imagery... the book really gets you thinking. In a book of this scope, it's very refreshing to find that the author has a gentle, conversational style and an open-minded approach towards the reader.

For example, she writes "Consider this a journey into possibility. I don't mind if you treat this as science fiction, science fact, an amusing tale... or purely just diverting balderdash... take it as you will and let it take you where it will."

... where it leads is to the I Ching, the ancient Chinese oracle that, according to Katya Walter, has: "unique fractal shorthand in a coded way that can merge physics and metaphysics."

Extensively referenced and full of diagrams—I really enjoyed this book and I'm looking forward to the next one in the series.

Mick Frankel
I Ching consultant-London, UK

-:::-

"This is the best book on this topic I've read and I've read a lot of them. Solid research on the science end without claiming unproven conclusions. The author simply explains her own TOE which she presents in a logical easy to understand manner. I appreciate her ability to speak to both the spiritual and scientific audience. Very thought provoking.

Winnie Hiller
Reader

-:::-

I think Katya Walter is a genius in that she can translate her right brain insights into left brain analysis with striking correlations and patient explanations. In this book, she's drawing on all her others to outline a sort of unifying theory of everything. Her discovery of the primordial pattern embedded in every level of creation, the "Master Code," is as significant as it sounds. Drawing on her first book, "The Tao of Chaos," she explains that this fractal pattern is originally created by the two primal pairs of opposites: space and time; matter and energy. She then follows the natural implications of that pattern to assert that there is a mirror opposite universe to ours of one-half dimensional space and three dimensional time. In her theory the missing or hidden parts of the pattern that we observe in physics (for example, the "arrow of time") are found in that mirror universe that she calls the Double Bubble universe."

Yeah, the "theory of everything" is a big assertion. Katya Walter's ideas are brave and bold- and impossible to prove. But, as a metaphysicist, she can't wait for the astro and nuclear physicists to catch up. Her books are sort of a field guide to physical reality for modern-day mystics. She explains her ideas through the models of biochemistry, a little math, geometry, and what she calls "the shorthand of the I Ching." She also includes her personal thoughts and dreams with her careful explanations of mathematics and physics. I'll admit, the mix takes getting used to. Yes, it's weird- but worth it!

Like any genius, the author is unconventional and eccentric, which could cause some people not to take her seriously. That would be a mistake, as a careful reading reveals an extremely intelligent and logical woman who asks the right questions. She simply doesn't stop asking, and may go a lot farther than most people are comfortable with, given that we may never have scientific proof for any of this. But, in this era of string theory which proposes many additional unknown dimensions, I wish the physicists would read her books. She could point them in the right direction, and may even save them some time with her simple and elegant theory of everything.

Erin Rose
Yoga Teacher & holistic Health Therapist

-:::-

Dr. Katya Walter's book *Double Bubble Universe* unites cutting-edge scientific research with her own inner 'deep see diving'. She accomplishes the incredible feat of inciting a paradigm-shift in the reader (to the realization that a love-intelligence underlies and pervades the physical universe), bringing her TOE to life! (unlike any other TOE I've read). Quantum mechanics, a physics of cold dead space, births 'quantum organics', a science of the fractal aliveness of the universe!

Dr. Walter skillfully creates an enjoyable, light read—provocative, funny, and digestible, dealing with perhaps the 'heaviest' topic of all—the structure and meaning of the creation and evolution of the universe. Read this book to witness the wedding of science and heart—watch how every whirling particle spins in the same wind as love's art! Who knows what could bubble up?!?

Peter Craig
Licensed Professional Counselor

I highly recommend this book for those who believe the current scientific paradigm is incomplete, and are looking for new explanations to fill the gaps. The *Double Bubble Universe* is a space-time, matter-energy, symmetry explanation of the physical universe. It introduces a scientific explanation of the physical laws of the universe with 20 questions. Written in layman's terms, Katya Walter's book encompasses the melding of Science and Metaphysics in which she intersperses and interweaves a personal dream with frontier science. Katya's writing skills are extensive and second to none, coining phrases that are truly inspired and unique.

Don Switlick
Institute for Neuroscience & Consciousness Studies

Replete with deep scientific insights that answer previously unanswerable questions yet accessible to lay readers, Ms. Walter's book offers the most comprehensive and useful T.O.E. ever. Comprehensive in that it not only explains reality from subatomic levels to the most macro perspectives, but it also links the physical and the spiritual and connects the evolution of the universe to the evolution of consciousness. Useful in that its elegant explanation of physical reality has implications that naturally lead one to contemplate how to live one's life more effortlessly and authentically. I highly recommend this book to all those truly thirsting to understand everything.

Kevin Blackwell
Stocks & Bonds Analyst

Have you ever wondered why we (human species), considering that which most concur is ineffable, continue effing it up? I've always thought having minds dead set on figuring things out in combination with phenomenon that exceed our ability to do so could be called 'God's dirty trick.' Dr. Walter has taken just such issues and playfully made a case worthy of consideration while mercifully maintaining the topic's ultimate ineffability. I found myself intellectually giggling throughout this read. Her consideration of the I Ching and our DNA alone is worth the purchase of her books.

MIL
Reader

I have read several other TOE books...the latest being Tom Campbell's My Big TOE. Wanted to read this one and see what new perspectives were explored. Was pleased to find that BOTH books stand on their own, and each adds new information without contradicting the other!! (So this is a "Must Read"!)

Great visual-inducing analogies and metaphors. Enjoyed reading about her personal background, leading to development of this book.

Descriptive down-to-earth language, even tho, on occasion, I have to look up a word in the dictionary!...which means you learn new concepts and words as you go.... The subject matter covers questions we all ask at one time or another...and the answers are creative and original...makes you think and gives perspective.

Hyphenated, descriptive words are used where needed, supporting the requirement for "hyphenated sciences" and new words to explain some of the more ephemeral aspects of mind and consciousness. Very insightful, creative application of recently-discovered fractal phenomena to explain its basic principles to everything in the universe. Plenty of references and web sites for the reader who wishes to explore further.

James Beal, Ph.D.
Aerospace & Electrical Engineer
-:::-

Katya Walter's series starting with the *Double Bubble Universe* integrates immense questions and insightful answers about the cosmos. She uses data, analogies, graphs, images, and stories that resolve together into one bubbling statement. A must read....

Rowena Pattee Kryder, Ph.D.
Dynamics & Foundations of Co-Creation
-:::-

Katya Walter is that rare writer who can merge so-called opposing systems, like science and metaphysics. For me, with a PhD in Literature and Communication, and a serious teacher of A Course In Miracles, and having had spiritual experiences myself, I am so grateful that she brings it all together, so I no longer have to wonder if I should trust those marvelous "intuitive" experiences enough to share them with others, without fear of ridicule. Just read Katya Walter if you think this is not possible. Thanks, Katya. We need you.

Helen Bonner, Ph.D.
Author

-:::-

Note from the Author

I FIND MANY MORE PEOPLE READ THIS BOOK THAN BOTHER TO REVIEW IT.

IF THIS BOOK WAS INTERESTING TO YOU, PLEASE REVIEW AND RATE IT.

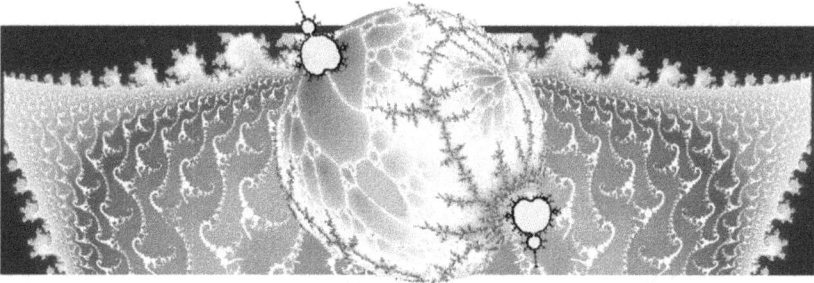

Bibliography

This is by no means all the books I consulted on writing this series, but here are what seemed to me most relevant to this volume.

Aczel, Amir D. *God's Equation: Einstein, Relativity, and the Expanding Universe.* New York: Four Walls Eight Windows. 1999.

Allgood, Kathleen T., Sauer, Tim, and Yorke, James A. *Chaos-an introduction to dynamical systems.* New York: Springer-Verlag. 1996.

Arrowsmith, William. "Speech of Chief Seattle, January 9, 1855." Arion 8: 461-64. 1969.

Bancroft, Anne. *Women in Search of the Sacred.* London: Penguin Books. 1996.

Barrow, John D. & Frank Tipler. *The Anthropic Cosmological Principle.* Oxford: Oxford University Press. 1988.

Bentov, Itzhak. *Stalking the Wild Pendulum: On the Mechanics of Consciousness.* New York: Bantam Books, 1981.

Bergmann, Peter G. *The Riddle of Gravitation.* New York: Charles Scribner's Sons, 1987.

Birken, Marcia, and Coon, Anne C. *Discovering Patterns in Mathematics and Poetry.* Amsterdam and New York: Rodopi Press. 2008.

Born, Max. *Einstein's Theory of Relativity.* New York: Dover Publications, 1965.

Buchanan, Keith. *China: The History, the Art, and the Science.* With Charles P. FitzGerald and Colin A. Ronan. New York: Crown Publishers. 1981.

Campbell, Joseph. *The Masks of God.* New York: Viking Press. 1964.

Capra, Fritjof. *The Tao of Physics.* New York: Bantam Books, 1977.

Carroll, L. Patrick, and Dychman, Katherine Marie. *Chaos or Creation.* New York: Paulist Press. 1986.

Carter, Brandon. "Large Number Coincidences and the Anthropic Principle in Cosmology," in *Confrontation of Cosmological Theories with Observation.* Longair, M.S., Editor. Dordrecht: Reidel. 1974.

Casti, John L. *Alternate Realities: Mathematical Models of Nature and Man.*

New York: John Wiley & Sons. 1989.

Chen, C. L. Philip, Tong Zhang, Long Chen, and Sik Chung Tam. *I-Ching Divination Evolutionary Algorithm and its Convergence Analysis.* Published in: IEEE Transactions on Cybernetics, Volume: 47, Issue: 1, Jan. 2017. Available in IEEE Xplore Digital Library at http://ieeexplore.ieee.org/document/7387702/

Coveney, Peter and Highfield, Roger. *The Arrow of Time: A Voyage Through Science to Solve Time's Greatest Mystery.* New York: Fawcett Columbine. 1990.

Cowen, Ron. "Loops of Gravity: Calculating a foamy quantum space-time" in *Science News*, Vol. 153. June 13, 1998.

D'Aquili, Eugene G. "Senses of Reality in Science and Religion: a Neuro-epistemological Perspective" in *Zygon*, Vol. 17, No. 4, December 1982.

Davies, Paul. *The Cosmic Blueprint.* London: Unwin Paperbacks. 1989.

...*Space and Time in the Modern Universe.* Cambridge: Cambridge University Press. 1977.

Davis, Philip and Hersh, Reuben. *The Mathematical Experience.* Boston: Houghton Mifflin. 1981.

DeWitt, B. and Graham, N, editors. *The Many-worlds Interpretation of Quantum Mechanics.* Princeton: Princeton University Press. 1973.

Doczi, György. *The Power of Limits.* Boulder,: Shambhala Publications. 1981.

Dyson, Freeman. *Infinite in All Directions.* New York: Harper & Row. 1988.

Einstein, Albert. *Relativity, the Special and General Theory: a Popular Exposition.* New York: Henry Holt & Company, 1921.

Ellis, George F.R. and Williams, Ruth M. *Flat and Curved Space-times.* Oxford: Clarendon Press. 1988.

Eddington, Arthur. *Fundamental Theory.* London: Cambridge University Press. 1946.

Fiedeler, Frank. Die Wende: *Ansatz einer genetischen Anthropologie nach dem System des I-ching.* Berlin: Kristkeitz. 1977.

Fraser, J.T., Lawrence, N., and Haber, F.C. *Time, Science, and Society in China and the West.* Volume V of The Study of Time. Amherst: The University of Massachuetts Press. 1986.

Fromm, Erich. The *Forgotten Language; An Introduction to the Understanding of Dreams, Fairy Tales and Myths.* New York: Grove Press, 1951.

Furtwangler, Albert. Answering Chief Seattle. Seattle: University of Washington Press. 1997.

Gardner, Martin. *The New Ambidextrous Universe: Symmetry and Asymmetry from Mirror Reflections to Superstrings.* Revised, updated edition. New

York: W.H. Freeman and Company.1990.

Gefter, Amanda. "Throwing Einstein for a Loop" in *Scientific American*, December, 2002.

Glass, Leon & Mackey, Michael. *From Clocks to Chaos: the Rhythms of Life*. Princeton: Princeton University Press. 1988.

Gleick, James. Chaos: *Making a New Science*. New York: Viking. 1987.

Gleiser, Marcelo. *The Dancing Universe: From Creation Myths to the Big Bang*. New York: Dutton. 1997.

Gopi Krishna. *Kundalini: The Evolutionary Energy in Man*. London: Stuart & Watkins. 1970.

Granet, Marcel. *La Pensé e chinoise*. Paris: Albin Michel. 1936.

Gribbin, John. *Genesis: The Origin of Man and the Universe*. New York: Delacorte Press. 1981.

...*In the Beginning: After COBE and Before the Big Bang*. Boston: Little, Brown and Company. 1993.

Hall, Edward T. *The Dance of Life: the Other Dimension of Time*. Garden City, N.Y.: Anchor Press. 1983.

Hawking, Stephen. *A Brief History of Time*. New York: Bantam. 1988.

Hobson, J. Allan, and McCarley, Robert W. "The Brain as a Dream State Generator" in *The American Journal of Psychiatry*, 134:12, December 1977.

Horgan, John. *The End of Science: Facing the Limits of Knowledge in the Twilight of the Scientific Age*. New York: Addison-Wesley Publishing, Helix Books. 1996.

Huajie Liu. "A Brief History of the Concept of Chaos." Peking University, Beijing, China. http://huajie.tripod.com/Paper/chaos.htm

Jenkins, R.C. *The Jesuits in China*. London. 1894.

Jones, Roger S. *Physics As Metaphor*. New York; Meridian, New American Library, 1982.

Jung, C.G. *Collected Works of C.G. Jung*. Princeton: Bollingen Press. 1959.

Kelso, J.A.S., Mandell, A.J., & Shlesinger, M.F. *Dynamic Patterns in Complex Systems—Conference Proceedings*. Teaneck, New Jersey: World Scientific. 1988.

Kucharski, Adam. *Forecasting the Chaos of Tornadoes*. http://theconversation. com

Laszlo, Ervin and Dennis, Kingsley. *Dawn of the Akashic Age: New Consciousness, Quantum Resonance, and the Future of the World*. Rochester, Vermont: Inner Traditions. 2013.

Leibniz, G. Wilhelm. *Explication de l'Arithmétique binaire, qui se sert des seuls caracteres 0 & 1.* 1703. Published in Memoires de l'Academie Royale des Sciences.

...*Zwei Briefe über das Binare Zahlensystem und die chinesische Philosophe.* Munich: Belser. 1968.

...the *Gottfried Wilhelm Leibniz Library* in Hanover, Germany has a compilation of many historical Leibniz-related documents. It is a rich source of information.

Li, T.Y. and Yorke, J.A. "Period Three Implies Chaos" in *The American Mathematical Monthly,* Vol. (10), 197 pp. 985-992. 1975.

Lindley, David. *The End of Physics: The Myth of a Unified Theory.* New York: Basic Books. 1993.

Lorenz, Edward. "Predictability: Does the Flap of a Butterfly's Wings in Brazil Set Off a Tornado in Texas?" Washington: American Association for the Advancement of Science. December 29, 1979.

...*The Essence of Chaos.* Washington: University of Washington Press. 1993.

Lovelock, James. *The Ages of Gaia.* New York: Norton. 1988.

McHarris, William C. *It from Bit from It from Bit...Nature and Nonlinear Logic.* Available at http://fqxi.org/data/essay-contest-files/McHarris_McHarrisFQXiPaper.pdf

Morris, Richard. *Dismantling the Universe: The Nature of Scientific Discovery.* New York: Simon and Schuster. 1983.

...*The Universe, the Eleventh Dimension, and Everything: What we Know and How We Know It.* New York: Four Walls Eight Windows. 1999.

Newmark, Joseph; Lake, Frances. Mathematics as a Second Language. New York: Addison-Wesley. 1982.

Noble, David F. *The Religion of Technology: The Divinity of Man and the Spirit of Invention.* New York: Alfred A. Knopf. 1997.

Pagels, Heinz. *The Cosmic Code: Quantum Physics as the Language of Nature.* New York: Simon and Schuster. 1982.

Parker, Barry. *Chaos in the Cosmos: The Stunning Complexity of the Universe.* New York: Plenum Press. 1996.

Penrose, Roger. *The Emperor's New Mind.* Oxford: Oxford University Press. 1989.

Pickover, Clifford A. *Black Holes: a Traveler's Guide.* New York: John Wiley & Sons, Inc. 1996.

Pratyagatmananda, S. *The Metaphysics of Physics.* Madras, India: Ganesh. 1964.

Prigogine, Ilya. *From Being to Becoming: Time and Complexity in the Physical Sciences.* San Francisco: W.H. Freeman & Company. 1980.

...*Order Out of Chaos: Man's New Dialogue With Nature*. With Isabelle Stengers. New York: Bantam Books. 1984.

Quin, Helen, and Witherell, Michael. "The Asymmetry between Matter and Antimatter" in *Scientific American*, October, 1998.

Raymo, Chet. *Skeptics and True Believers: The Exhilarating Connection Between Science and Religion*. New York: Walker and Company. 1998.

Roszak, Theodore. *The Gendered Atom: Reflections on the Sexual Psychology of Science*. Berkeley: Conari Press. 1999.

Russell, Bertrand. *Introduction to Mathematical Philosophy*. London: Allen & Unwin. 1919.

Ryan, Robert E. *The Strong Eye of Shamanism*. Rochester, Vermont: Inner Traditions International. 1999.

Schonberger, Martin. *The I Ching and the Genetic Code*: the Hidden Key to Life. New York: ASI Publishers. 1979.

Sessions, Roger. *Wisdom's Way: The Christian I Ching*. The Christian I Ching Society of Cedar Park, Texas. 2015.

Shchutskii, Iulian. *Researches on the I Ching*. London: Routledge & Kegan Paul. 1980.

Sheldrake, Rupert. *A New Science of Life*. Los Angeles: Tarcher. 1981.

Shubnikov, A.V., and Koptsik, V.A.; trans. editor, Harker, David. *Symmetry in Science and Art*. New York: Plenum Press, 1974.

Smith, Richard J. *The I Ching: A Biography*. Princeton: Princeton University Press. 2012.

Smith, Steven W. *The Scientist and Engineer's Guide to Digital Signal Processing*. San Diego, CA: California Technical Publishing. 1998.

Sudarshan, George, and Rothman, Tony. *Doubt And Certainty: The Celebrated Academy Debates on Science, Mysticism, Reality*. Basic Books. 1998.

Susskind, Leonard. "Black Holes and the Information Paradox" in *Scientific American*, April, 1997.

Szamosi, Geza. *The Twin Dimensions: Inventing Time & Space*. New York: McGraw-Hill Book Company. 1986.

Tiller, William. *Conscious Acts of Creation: the emergence of a new physics*. Walnut Creek, California: Pavior Publishing. 2001.

...*Science and Human Transformation: Subtle Energies, Intentionality, and Consciousness*. Walnut Creek, California: Pavior Publishing. 1997.

Turner, Michael S. "Quarks and the Cosmos" in *Science*, Vol. 315, January 5, 2007.

Von Franz, Marie-Louise. *Number & Time*. Evanston: Northwestern University Press. 1974.

Weinberg, Steven. *Gravitation and Cosmology: Principles and Applications of the General Theory of Relativity*. New York: John Wiley and Sons, 1972.

...*The First Three Minutes: A Modern View of the Origin of the Universe (updated)*. New York: Basic Books, 1988.

...*Dreams of a Final Theory: The Search for the Fundamental Laws of Nature*. New York: Pantheon Books. 1993.

Wheeler, John A. *Geons, Black Holes and Quantum Foam: A Life in Physics*. New York: Norton and Company. 1998.

Wiggins, Stephen. *Global Bifurcations and Chaos: Analytical Methods*. New York-Berlin: Springer-Verlag. 1988.

Weisstein, Eric W. "Klein Bottle." From MathWorld--A Wolfram Web Resource. http://mathworld.wolfram.com/KleinBottle.html

Wigner, Eugene. *Symmetries and Reflections*. Westport, Conn.: Greenwood Press. 1967.

Williams, Garnett P. Williams. *Chaos Theory Tamed*. London: Taylor & Francis. 1997.

www.ingramcontent.com/pod-product-compliance
Lightning Source LLC
Chambersburg PA
CBHW071551200326
41519CB00021BB/6701